Aircraft Mechanic and Service Technician: A Career Guide

Aircraft Mechanic and Service Technician: A Career Guide

Patrick Grant

STATES
ACADEMIC PRESS
www.statesacademicpress.com

Aircraft Mechanic and Service Technician: A Career Guide
Patrick Grant
ISBN: 978-1-63989-808-4 (Hardback)

Published by States Academic Press,
109 South 5th Street,
Brooklyn, NY 11249, USA

Visit us on the World Wide Web at:
www.statesacademicpress.com

Cataloging-in-Publication Data

Aircraft mechanic and service technician : a career guide / Patrick Grant.
p. cm.
Includes bibliographical references and index.
ISBN 978-1-63989-808-4
1. Airplanes--Maintenance and repair--Vocational guidance. 2. Aviation mechanics (Persons)--Vocational guidance.
I. Grant, Patrick.
TL671.9 .A37 2023
629.134 6--dc23

Table of Contents

Preface

Aircraft mechanics and service technicians are professionals who carry out aircraft maintenance. They are entrusted with the inspection, repair and supervision of aircraft and aircraft systems. They are expected to have a good understanding of hydraulics, pneumatics, weight and balance, electricity, basic physics, ground operations of aircraft, cleaning and corrosion control, etc. Knowledge of rules and regulations, navigation and warning systems, position and warning systems, fire protection systems, cabin atmosphere control systems, etc. are also vital to the work of an aircraft mechanic and service technician. This textbook provides comprehensive insights into the duties and responsibilities of aircraft mechanics and service technicians. It aims to serve also as a career guide to students interested to work in these domains. Those in search of knowledge to broaden their understanding of this field will find this book immensely helpful.

A detailed account of the significant topics covered in this book is provided below:

Chapter 1- Aircraft mechanics and service technicians are professionals who carry out various tasks required for the smooth functioning of aircraft and aircraft systems, such as inspection and supervision, maintenance and preventive maintenance. This is an introductory chapter, which discusses the roles and responsibilities of aircraft mechanics and service technicians. It also explores the basics of aircraft maintenance and aircraft repair and the ways of getting trained in aviation repair, becoming a helicopter mechanic or an FCC licensed avionics technician, etc.

Chapter 2- The airframe of an aircraft refers to its mechanical structure consisting of the wings, fuselage and undercarriage. Airframe technicians are tasked with the inspection of aircraft frames, mechanical and electrical systems for detecting defects, wear and other issues in the airframe. They are also responsible for the repair and replacement of parts. This chapter discusses the vitals of repairing minor aircraft skin damage and fueling a plane, as well as the ways to become a successful airframe mechanic.

Chapter 3- Avionics refer to the electronic systems used in aircraft. It encompasses navigation systems, communications systems, monitoring systems, flight recorders, aircraft flight-control systems, aircraft management systems, etc. An avionics technician is specialized in the maintenance of all these electronic components and the installation of autopilots, radios and passenger entertainment systems. The aim of this chapter is to explore the various aspects of radio communication, aircraft weight and balance, autopilot systems, antennas, etc. and the various responsibilities of avionics technicians.

Chapter 4- An aircraft powerplant is a power source such as an engine that helps airplanes and helicopters propel off the ground. The professionals entrusted with the maintenance of these engines are known as powerplant mechanics. The chapter elucidates the ways to become an aircraft engine specialist and aircraft powerplant technician.

It gives me an immense pleasure to thank our entire team for their efforts. Finally in the end, I would like to thank my family and colleagues who have been a great source of inspiration and support.

Patrick Grant

Chapter 1

Aircraft Mechanic and Service Technician: Roles and Responsibilities

Aircraft mechanics and service technicians are professionals who carry out various tasks required for the smooth functioning of aircraft and aircraft systems, such as inspection and supervision, maintenance and preventive maintenance. This is an introductory chapter, which discusses the roles and responsibilities of aircraft mechanics and service technicians. It also explores the basics of aircraft maintenance and aircraft repair and the ways of getting trained in aviation repair, becoming a helicopter mechanic or an FCC licensed avionics technician, etc.

Aircraft Mechanic

Aircraft mechanics maintain and repair the avionic and mechanical equipment needed to make planes and helicopters operate safely. This includes repairing, inspecting and maintaining the structural, mechanical and electronic elements of the aircraft, replacing parts, measuring wear and assessing whether parts need repair or replacement. Aircraft mechanics use a wide variety of diagnostic devices, computers, hand and mechanical tools to assess and fix aircraft, and maintain aircraft standards as required by the Federal Aviation Administration (FAA). Mechanics can specialize in a specific form of aircraft or be a generalist, and most will have an FAA Airframe and Powerplant (A&P) certificate.

How to Become an Aircraft Mechanic

The route to become an aircraft mechanic requires specialized training, followed with certification by the Federal Aviation Administration (FAA). To become certified, you'll need 18 to 30 months of practical work experience, and you'll have to pass practical and written exams. You can work in the armed forces as an aircraft mechanic without a license, but once you leave the service you'll need to obtain FAA certification unless you plan to work under direct supervision of a licensed aircraft mechanic.

Method 1

Acquiring Work Experience as a Civilian

1. Go to school. There are 170 Aviation Maintenance Technician schools in the US, that are approved by the Federal Aviation Administration (FAA). Programs usually last 12 to 24 months, and most require a high school degree or a GED.

- Graduation qualifies you to sit for the FAA exams required for certification.
- Graduates often earn higher salaries than mechanics who get their experience through apprenticeship or the military.

2. Work as an apprentice. If you do not have a mechanic's license from the Federal Aviation Administration (FAA), you can still work on avionics equipment in a limited capacity. You'll need 18 months of practical, on-the-job experience for each type of certificate (power plant or airframes), or 30 months total to get both certificates.

- You will be allowed to work only under direct supervision of someone who has an FAA mechanic's license.
- You'll be required to produce proof of the work you did – pay stubs, and a letter from your employer – before you can take the FAA tests.
- You will not be authorized to approve planes, engines, appliances, or parts to be returned to service. Someone with a license will need to sign off on that.
- Without a license, you will be restricted to basic mechanical work, and will not advance to higher levels in the business.

3. Prepare yourself for a challenging work environment. Avionics mechanics can earn very good

salaries, and the work tends to be interesting and varied. However, there are pros and cons that you should be aware of if you are considering this career path.

- You'll be using sophisticated equipment and potentially dangerous power tools.
- New mechanics are usually expected to work weekends and nights, because work is needed 24 hours a day, 7 days a week. The better shifts go to mechanics with seniority.
- You'll be working in environments where the noise levels are extremely high.
- The work is very physical and involves climbing, crawling, and carrying heavy parts.
- You will work against tight deadlines – the planes need to be ready on time, every time. This can be very stressful for some people.

Method 2

Acquiring Work Experience through the Armed Forces

1. Choose an approved job speciality. Your practical experience in the military is only applicable to certification if it's one of the specific jobs that has been approved by the Federal Aviation Administration (FAA). You'll need to contact your local FAA Flight Standards District Office (FSDO) for a list of these jobs, because the list changes without notice. Your military recruiter should also be familiar with specific job listings for FAA certification requirements.

- Each branch of the military has its own job codes. There are more than 50 MOS codes for the Air Force that describe jobs that are reported as being FAA-approved work experience.
- The US Army has around 20 specific MOS job codes that qualify, including: 15N10/30 (Avionics Mechanic, Airframe experience), 15M10/30 (UH-1 Helicopter Repairer, Airframe and Powerplant experience), and 15B10/30 (Aircraft Powerplant Repairer, Powerplant experience).

2. Contact a military recruiter. If you believe you meet the basic requirements for US military eligibility, you can get in touch with your local recruitment office at military.com. Basic military requirements are:

- Be a US citizen or resident alien.
- You'll need to be at least 18, or 17 with permission from your legal guardian.
- Most jobs require a high school diploma (there are a few that will allow a GED, but not many.)
- You'll need to be in good physical and mental condition, and pass medical examinations.
- You will need to pass two separate drug tests.
- For the Air Force, you must also have an AFQT score of at least 50, be between 17 and 27 years old, and have no more than two dependents.
- For the Army, you must be between 17 and 34, have no more than two dependents, and have a minimum AFQT score of 31.
- For the Marines, you must be between the ages of 17 and 29, and have a minimum score of 32 on the AFQT. Women are allowed to serve in all jobs except combat positions.
- For the Navy, you must be between 17 an 34, and have a minimum AFQT score of 50. Women are allowed to sign up for any job except the Navy Seals and positions on submarines.

3. Work as an aircraft mechanic in the military. Jobs vary in each branch of the US military, so

you will need to speak to a recruiter to get specific details about aircraft mechanic positions. Each branch of the military has its own basic requirements for enlisting.

- Certain occupations have additional requirements beyond the basics, such as a certain score on the Armed Services Vocational Aptitude Battery (ASVAB) test.
- You may need to demonstrate aptitude in math, science, and technology to convince a recruiter you'd do well as an avionics mechanic (through test scores or your school transcript).

4. Get an official letter from your superior. When you leave the service, you'll need to get a letter from your boss listing the types of aircraft and engines you worked on, and how many hours of practical experience you received. Training time does not count toward practical experience hours, only actual hours worked.

Method 3

Obtaining Aircraft Mechanic Certification

1. Pass the oral and practical examinations. Once you have obtained the required amount of practical training, you will need to pass three tests administered by the Federal Aviation Administration (FAA).

- A Designated Mechanic Examiner will test your practical skills.
- The Examiner will also administer an oral exam – you'll answer his questions in person, while he takes notes.
- There are 43 different technical topics that you'll need to be familiar with in order to pass these tests.
- For a single certificate exam (airframe or power plant), expect the test to last about 8 hours.

2. Pass the written exam. You will need to go to your local Federal Aviation Administration (FAA) office and submit proof that you have all the required experience for certification. If the FAA inspector determines that you are eligible to take the written exam, you can schedule your exam at one of the approved computer testing centers. There are testing centers in most large towns and cities in the US.

- Study very carefully for your written exam. If you fail any part of the test, you'll need to wait at least 30 days before taking it again.
- The 30-day wait requirement may be waived if you can show proof that you've gotten additional training in the area you failed.
- Every section of the written exam must be passed within 24 months of your first attempt.

3. Get a mechanic's license from the Federal Aviation Administration (FAA). You can get a license

in a power plant rating, an airframe rating, or both. you are a US citizen and want to work as an aircraft mechanic in the United States, you must meet certain basic standards. You will need a mechanic's certificate to work as an aircraft mechanic.

- You must be 18 years of age, or older.
- You must be able to speak, write, read, and comprehend English.
- You must have 18 months of practical experience with airframes or power plants, or 30 months of practical experience working on both power plants and airframes at the same time.
- You must pass the required practical, oral, and written exams.

4. Get a mechanic's license from the Federal Aviation Administration (FAA) requirement if you are not a US citizen. If you aren't a US citizen, you can still qualify for a mechanic's certificate if you meet all of the US citizen standards, along with some additional requirements.

- You'll need to prove that you need a mechanic's certificate to work on US-registered civil aircraft.
- You will need to show your non-US passport to the examiner.
- Your employer will need to write a letter explaining exactly what aircraft mechanical work you performed for him, and for how long.
- You will need a letter from the foreign equivalent of the FAA listing your qualifying work experience, or a letter to this effect from the International Civil Aviation Organization (ICAO).
- All papers must be dated and signed, and must be the original document – no copies are allowed.
- You'll need to pay a fee to have your documents reviewed by FAA officials.
- English-language requirements are occasionally waived if you will not be working within the US. In this situation, your license would limit you to work outside the US.

5. Get a repairman certificate, if you are already on the job. The Federal Aviation Administration (FAA) also issues repairman certificates, which restrict the bearer to specific tasks (such as repairing a specific type of engine or engine component, for a specific type of aircraft). The work is performed for commercial operators, FAA-approved repair stations, or air carriers that have been approved to do this kind of work. The requirements for a repairman certificate are:

- Your employer will need to recommend you for a repairman certificate before you can apply.
- You'll need to be 18 or older.
- You must be able to write, read, understand, and speak the English language.
- You will need work experience that qualifies you to perform maintenance on aircraft or parts.
- You must already be employed on a specific job that requires qualifications by an FAA-certified Repair Station, air carrier, or commercial operator.
- You will need 18 months of on-the-job experience in the specific job that you will be doing with your repairman certificate, or you will need to complete a school training course at an FAA-approved facility.

How to get Licensed as an Aircraft Mechanic

The Federal Aviation Administration (FAA) issues aircraft mechanic's licenses in the United States. You will need a license from the FAA even if you work outside the United States, provided you work on civil aircraft that is registered in the United States. To get the license, you must pass a series of tests after gaining sufficient experience or training.

Part 1

Checking if You Satisfy Requirements

1. Meet the age requirements. You must be at least 18 years old in order to qualify for an aircraft mechanic's license. You will also need over a year of practical experience, so you realistically might not be able to apply until you are over 18.

2. Understand English. You also need to be able to read, write, speak, and understand the English language. However, if you live outside the United States, then the FAA may waive the English language requirement.

- If waived, your certificate will only be valid outside the United States.

3. Check the requirements if you live outside the U.S. Non-citizens can also get certificates. They

must be at least 18 years old and pass the same types of tests. In addition, non-citizens must meet the following requirements:

- Demonstrate that you need a mechanic certificate in order to maintain civil aircraft that is registered in the U.S.
- Show that you are not a U.S. citizen or a resident alien in the U.S.
- Get a detailed statement from your employer explaining what specific types of maintenance you perform on each aircraft and the length of time you have performed it.
- Get a letter from the airworthiness authority in the country where you got experience, or get a letter from the advisor of the International Civil Aviation Organization (ICA) which validates your maintenance experience.
- Pay a fee for document review.
- Show the examiner your passport.

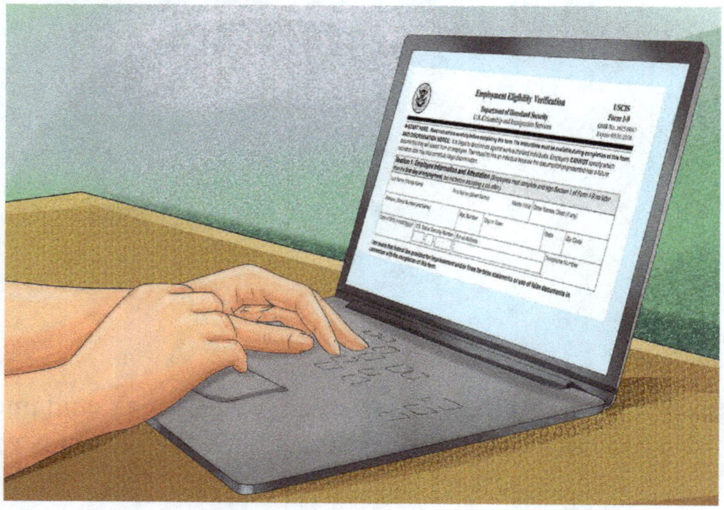

Part 2

Gaining Sufficient Experience

1. Find an FAA-Approved Aviation Maintenance Technician School. You can search a database on the FAA website to find schools. You can search by city, state, and country. Once you have located a nearby school, you should contact them and ask for information.

- You usually need a high school diploma or General Education Diploma (GED) in order to get into most schools.
- School usually lasts from 12-24 months.
- Graduates usually get higher starting salaries than those who don't graduate from a technical school, so you should seriously consider attending.

2. Get practical experience instead. If you don't want to go to a technical school, you can get 18 months of practical experience with airframes. If you want a power plant certificate as well, then you could get 30 months of experience that covers both airframes and power plants.

- Many people get their practical experience with on-the-job training in the military, though you can also get civilian on-the-job training.

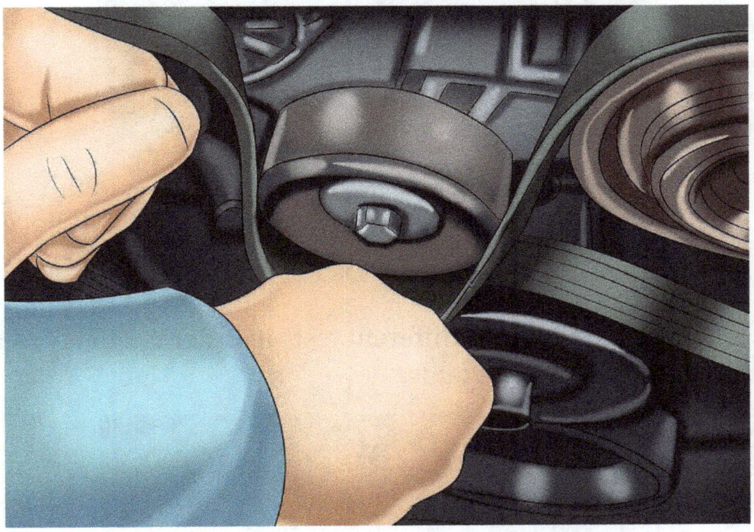

3. Document your practical experience. You should get pay receipts and a log book signed by your supervising mechanic. You should also get a notarized statement from your employer as to your experience.

- If you get military experience, then you can count only the time you got working in the specialty (not training for it). You should obtain an official letter from your military employer certifying your length of service, as well as the amount of time you worked in each specialty. The letter should also state the make and model of the aircraft or engine on which you got practical experience and where you got that experience.

4. Have an interview. If you are relying on practical experience, then you will have to interview with the FAA Airworthiness inspector. This person will review your paperwork and credit your work experience after a satisfactory interview.

Part 3

Taking Required Examinations

1. Identify the tests. You need to take three different tests in order to qualify as an aircraft mechanic. There are oral, practical, and written exams.

2. Register for the oral and practical exams. A Designated Mechanic Examiner will administer the

oral and practical tests. You can get a list of examiners from your local FAA office. The tests last for about eight hours and cover 43 different technical subjects.

3. Register for a written test. You can register by showing your proof of experience to an FAA inspector at the local FAA office. There are different tests for airframe and power plant mechanic certificates and a general test that covers both. Based on your experience, the FAA inspector will decide if you meet the requirements to take one of the tests.

- You will then register an appointment to take the test at a computer testing facility, which are located worldwide. The FAA has a list of testing centers at their website.
- To register, you will need to have an acceptable form of identification, which should include a recent photograph, signature, and actual residential address. For example, you could use a drivers' license, passport, or military identification card.

4. Study for the written test. You can get sample questions for the general, airframe, and power plant tests. The sample questions are available in PDF format at the FAA website.

5. Pass all exams. You must take and pass all three exams in a 24-month period. After passing, FAA will issue your certificate to you.

- If you fail a test, then you have to wait 30 days before taking it again. However, you can take it earlier if you give a letter to the Examiner stating that you have gained additional training in the area you failed.

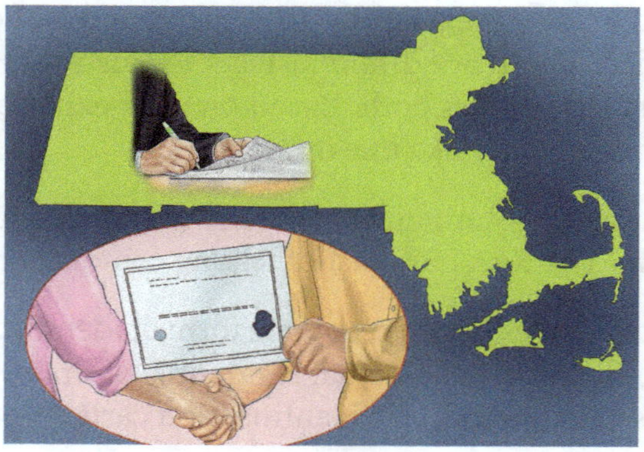

How to get Aviation Repair Career Training

Aviation repair workers engage in a broad range of duties that includes aircraft mechanical repair work, preventative maintenance work as well as aircraft inspections. The work environment for those in aviation repair can be either indoors within aircraft hangers and work stations or outside in the areas where aircraft are kept. Opportunities for aviation repair jobs can be found at both large airlines as well as smaller commuter airports. Educational requirements for aviation repair positions entail extensive training and for some positions such as maintenance work, certification and licensure is required.

Steps

1. Conduct some research on aviation repair career training to learn more about the trade to see if it is the right choice for you. Positions within this field require that workers have a strong mechanical aptitude.

2. Evaluate your ability to work with complex machinery and to accurately diagnose and repair mechanical systems. Aviation repair positions involve working with advanced electronics and technical manuals.

3. Decide if you want to get training within a specialized area of aviation repair. Aviation repair workers can focus their training on areas such as airframe mechanics, avionics equipment repair, or hydraulics and electrical systems repair, to mention a few.

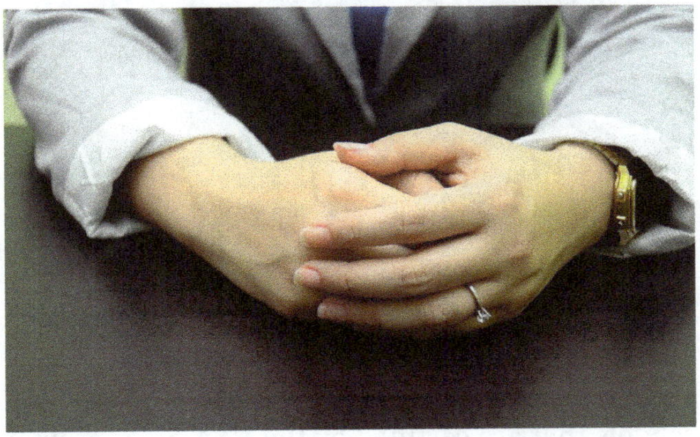

4. Realize that if you decide to become a preventative maintenance mechanic in the aircraft industry, the Federal Aviation Administration (FAA) requires that workers are to be licensed and certified.

5. Choose an educational training program that is approved and certified by the FAA. The FAA's website offers information, resources and guides for those exploring the possibility of an aviation repair career.

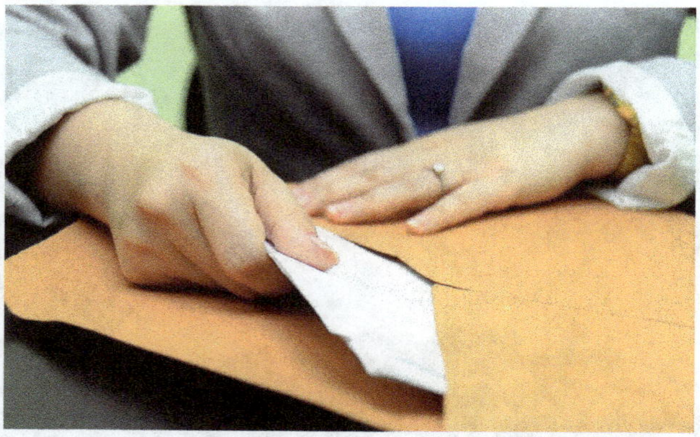

6. Weigh the various aviation repair career training program options available to you. Some training programs can be completed in 12 months. There are also associate's and bachelor's degree programs available if you want to pursue a more in-depth level of training and education.

7. Recognize that those schools that are certified by the FAA must provide their students with instruction and training amounting to a minimum of 1,900 class hours. Good programs will include hands-on training similar to the actual work performed out in the field when performing aircraft repairs.

8. Pick several aviation repair career training programs that meet your career goals. Research each program further to learn more about the options available.

9. Narrow your choices down to a select few aviation repair career training programs. Depending on admission requirements and openings, you may want to apply to more than one school.

10. Complete the application forms for admission to the aviation repair career training programs that you are interested in. Upon successful acceptance, you will then be prepared to commence your studies in the field of aviation repair training.

11. Include coursework in subjects such as computer science, mathematics, electronics and mechanical drawing to broaden the depth of your learning as it directly relates to the principles of aviation mechanics. Strong writing and reading skills are equally important.

12. Develop your skills and advance your knowledge in the field of aviation repair training and you will soon be on your way to a new career.

Service Technician

An automotive service technician (or auto mechanic) is someone who inspects, maintains, and repairs electric, gas, hybrid and alternative fuel vehicles. In smaller businesses, their duties may include a full range of repair and maintenance services. In larger businesses, they may specialize in certain areas, such as brakes, transmissions or air conditioning systems.

Automotive service technicians must discuss automotive problems—along with options to fix them—with their customers. Automotive businesses depend on repeat clients, therefore a technician must always be courteous, a good listener, and ready to answer customers' questions.

Duties of a Service Technician

Service technicians typically do the following:

- Test parts and systems to ensure they are working properly
- Identify mechanical problems, often by using computerized diagnostic equipment
- Follow checklists to ensure that all critical parts are examined
- Test and lubricate the vehicle's engine and other major components
- Perform basic care and maintenance, including oil changes, tune-ups, and tire rotations
- Repair or replace worn parts, such as brake pads and wheel bearings
- Disassemble and reassemble parts
- Use testing equipment to ensure that repairs and maintenance are effective
- Explain to clients their automotive problems and the repairs done on their vehicles

Automotive service technicians work on traditional mechanical components, such as engines, transmissions, belts, and hoses. However, they must also be familiar with a growing number of electronic systems. Braking, transmission, and steering systems, for example, are controlled primarily by computers and electronic components. Other integrated electronic systems, such as accident-avoidance sensors, are becoming common as well. In addition, a growing number of technicians are required to work on vehicles that run on alternative fuels, such as ethanol and electricity. Recreational vehicle service technicians specifically work on RV's.

Service technicians use many different tools, including computerized diagnostic tools and power tools such as pneumatic wrenches, lathes, welding torches, and jacks and hoists. These tools usually are owned by their employers. Service technicians also use many common hand tools, such as pliers, wrenches, and screwdrivers, which generally are their own. In fact, experienced workers often have thousands of dollars invested in their personal tool collection.

Automotive service technicians sometimes specialize in a particular type of repair that may be subject to specific regulations or procedures. For instance, those focused on air-conditioning system repairs must follow federal and state regulations governing the handling, recycling, and disposal of refrigerants.

In some shops, automotive service technicians may specialize. The following are types of service technicians:

Automotive Air-Conditioning Repairers

- Install and repair air conditioners and service parts, such as compressors, condensers, and controls. They are trained in government regulations related to their work.

Brake Repairers

- Adjust brakes, replace brake linings and pads, and make other repairs on brake systems. Some technicians specialize in both brake and front-end work.

Front-End Mechanics

- Align and balance wheels and repair steering mechanisms and suspension systems. They frequently use special alignment equipment and wheel-balancing machines.

Transmission Technicians and Rebuilders

- Work on gear trains, couplings, hydraulic pumps, and other parts of transmissions. Extensive knowledge of computer controls, the ability to diagnose electrical and hydraulic problems and other specialized skills are needed to work on these complex components.

Tune-Up Technicians - adjust ignition timing and valves and adjust or replace spark plugs and other parts to ensure efficient engine performance. They often use electronic testing equipment to isolate and adjust malfunctions in fuel, ignition, and emissions control systems.

Workplace of a Service Technician

Most automotive service technicians work full time for private companies, and about 18% are self-employed. Most work in well-ventilated and well-lit repair shops. Although many problems can be identified and fixed with computers, technicians frequently work with greasy parts and tools, sometimes in uncomfortable positions.

How to Become an FCC Licensed Avionics Technician

Avionics technicians are responsible for performing a variety of tasks that may include maintenance and repair work on aircraft navigation and radio communications as well as weather radar and missile control systems. Work activities encompass numerous tasks such as inspection and installation, adjustments, service and repair. An Avionics Technician License is not presently offered through The FCC (Federal Communications Commission) but they do currently issue several types of commercial operator licenses. The General Radiotelephone Operators License is required by the FCC for those doing repair and maintenance work on FCC licensed radiotelephone aviation transmitters.

Steps

1. Research the field of avionics to learn more about the work to see if the field is a good match for you. The work that avionics technicians engage in is highly specialized and requires extensive training as well as certification and licensing.

2. Visit the website of the FCC to review the description of General Radiotelephone Operators License and to get an overview of the examination. Requirements include being a legal resident of the United States, being able to both receive and transmit messages in English and passing the examination.

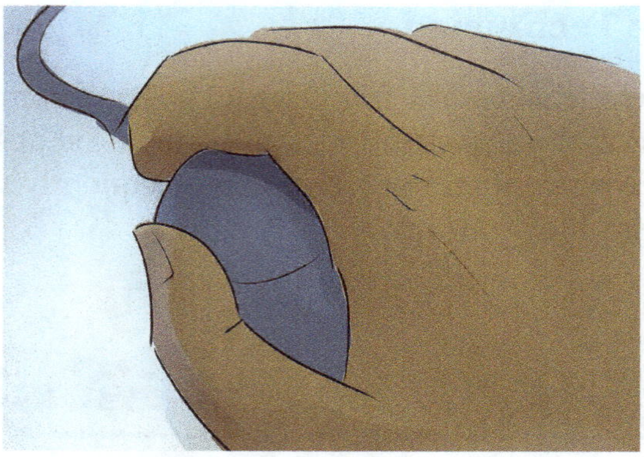

3. Browse the Internet and explore various programs that provide technical training in preparation for a position as an avionics technician if you have no experience within the field. Obtaining an education in avionics lays the necessary groundwork prior to completing the examination for the General Radiotelephone Operators License.

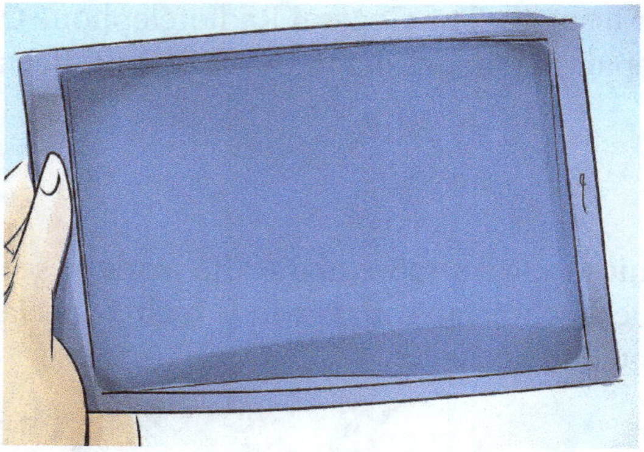

4. Consider taking some refresher avionics coursework if you are already working in the field but want to become certified and licensed. Some preparatory study and tutorials will increase your chances of passing the FCC's General Radiotelephone Operators License Examination so that you can advance in your career.

5. Review the various sections of the written examination for the General Radiotelephone Operators License available on the website of the FCC. Successfully passing this examination requires testing under Element 1 and Element 3.

6. Review the portion that focuses on Written Element 1 of the General Radiotelephone Operators License on the FCC website. This portion focuses on basic operating practice and radio law and requires correctly answering a minimum of 18 questions out of a total of 24 questions.

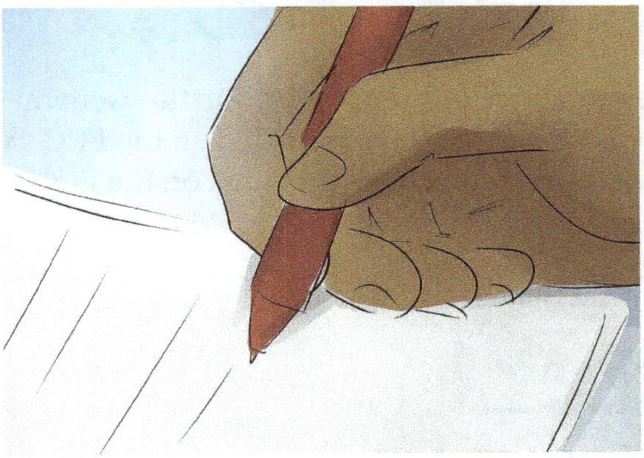

7. Read through the overview provided for Written Element 3 of the General Radiotelephone Operators License on the FCC website. This topic covers general radiotelephone and electronic fundamentals necessary to complete repairs, adjustments and maintenance on transmitters for both receivers and radios. Applicants must successfully respond to a minimum of 75 out of 100 questions to pass this portion.

8. Browse through the FCC's website to discover that sample question and answer sheets are

available for perspective examinees to acquire in preparation for the General Radiotelephone Operators License Examination.

9. Realize that study guides to assist in preparation for the General Radiotelephone Operators License Examination can be obtained via links provided on the FCC's website. There is additional information available for interested individuals provided on the FCC's site for examination schedules, locations and fees.

Helicopter Mechanic

Helicopter mechanics work in hangars or on runways with simple tools such as wrenches and hammers, as well as more refined diagnostic equipment. Candidates for this career must have knowledge of tools and be able to work under pressure and in various weather conditions. They generally train through programs at aviation maintenance technology schools. Like airplane mechanics and service technicians, helicopter mechanics are required to obtain certification from the Federal Aviation Association (FAA).

Required Education	FAA-certified training program
Certification	FAA certification required
Projected Job Growth (2014-2024)*	1% for aircraft mechanics and service technicians
Median Salary (2015)*	$58,370 for aircraft mechanics and service technicians

Salary of a Helicopter Mechanic

According to the U.S. Bureau of Labor Statistics (BLS) in 2015, most aircraft mechanics and service technicians, including helicopter mechanics, made anywhere from $16.71 per hour or less for the lowest-paid 10% to $42.81 per hour or more for the top-paid 10%. This is between $34,770 and $89,050 annually. The median average annual salary of aircraft mechanics and service technicians, including helicopter mechanics, was $58,370.

Mechanics who work in the scheduled air transportation industry tend to earn more those working in the nonscheduled air transportation industry, but competition for employment can be fierce. Another factor that affects compensation is geographical location. The BLS states that higher salaries are paid to individuals working in certain states, including Kentucky, New Jersey and Washington.

Job Duties

Helicopter mechanics conduct inspections of aircraft according to specific regulations. They use instruments to measure parts in order to determine if they are worn and need replacement. Mechanics work with many different types of tools including hammers, lifts, metal cutters and punches. They also are responsible for some of the technological aspects of the aircraft and must use various software programs to diagnose the problem a particular craft may be having. Helicopter mechanics must have fine-tuned troubleshooting and problem-solving skills.

Certification Requirements

There are two types of certification offered to aviation mechanics through the Federal Aviation Association (FAA): airframe mechanics and powerplant mechanics. Many employers prefer to hire someone who has a combined A&P certificate. The FAA requires candidates to complete 18 months of work experience before they are eligible to apply for certification and 30 months if they are attempting a combined powerplant and mechanics certificate.

Most helicopter mechanics learn the necessary skills from an aviation maintenance technology school that is certified by the FAA. As of 1992, the FAA requires that a minimum of 1,900 hours of coursework be offered at certified mechanic schools training individuals for combined certification. Only 1,140 hours are needed to train students in either airframe or powerplant maintenance. In order to receive any certification, mechanics must pass written and oral exams to demonstrate their skills.

Helicopter mechanics are largely responsible for inspecting helicopters and other aircraft to ensure they are ready for flight and meet specific regulations. A helicopter mechanic must have excellent problem-solving and computer skills, in addition to a solid understanding of mechanics. Completion of an FAA-certified training program is required for this career, and job growth is expected to be minimal over the 2014-2024 decade.

How to Become a Helicopter Mechanic

Are you interested in becoming an a helicopter mechanic? Helicopter mechanics enjoy employment opportunities all over the world. Keeping a helicopter airworthy takes a skilled technician. Helicopters, by design, require much more maintenance than their fixed wing counterparts.

Steps

1. Becoming a helicopter mechanic is the same process as becoming a fixed wing aircraft mechanic. In the United States, this means getting the airframe and power plant certificates, or the A&P license as it is commonly called.

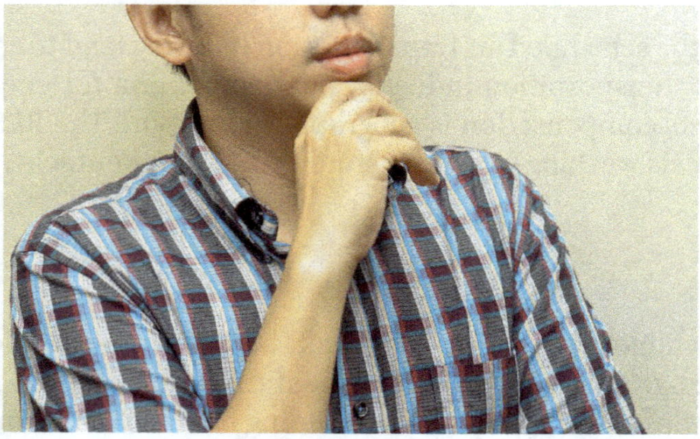

2. To get the A&P license, one must either attend an aviation mechanic school or have 30 months of hands on experience working on aircraft. This means working on the various systems of the aircraft and not just one area. Aircraft mechanic schools typically last 2 years or so. Military members who maintained aircraft in the service can use that towards the 30 months of required experience. Keep in mind that not all MOS's count towards both airframe and power plant.

3. Once the 30 months experience requirement is met or one has attended an aviation mechanic school, it is then possible to test for the A&P license. The tests are broken down into 3 tests and 3 sections to each test. General, airframe and power plant. Each one of those sections have a written, oral and practical test. To pass, one must get at least 70% in each section.

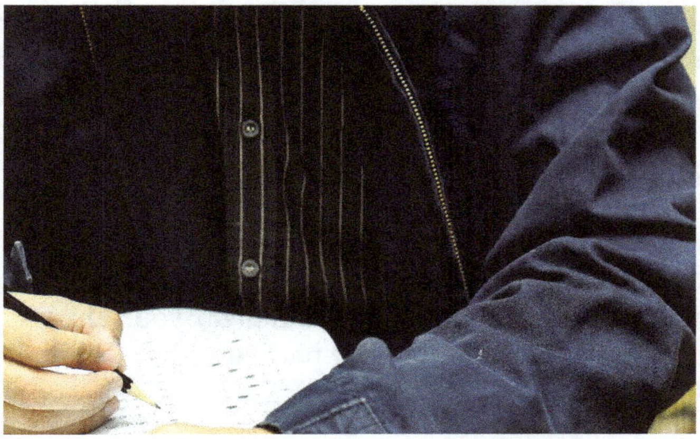

4. After acquiring the A&P license, one can now apply for helicopter mechanic positions.

Tips for Repairing Airplane Damage

A damaged aircraft can be disheartening and maybe even scary if the damage occurred while in flight. Getting an aircraft repaired is a little different than getting an automobile repaired. It's important to understand how to get your aircraft repaired quickly while following the proper procedures so that you can be back in the air as quickly and safely as possible.

File An Aircraft Insurance Claim

Aircraft insurance may cover the costs of repairs, so filing a claim as soon as possible may save you money. When filing the claim, you will need a copy of the registration, a copy of the logbook pages showing the date of the most recent inspection, and information about the pilot if the damage occurred while in flight. Photographs of the damage may also help with the insurance assessment.

Call an FAA Certified Mechanic

FAA certified mechanics are often available on call. Since an airplane is not as easy to move as a car, many aircraft mechanics will travel to damaged planes, even on nights and weekends. This may be your best option if you notice damage during a pre-flight inspection or if damage is spotted during an inspection. The ability levels of mechanics may vary, so there is a possibility that the first mechanic that you deal with my refer you to another mechanic if the damage is extensive or complicated to fix.

Have the Damage Repaired at the Airport

Airports have mechanics on site that are able to at least do minor repairs. If your aircraft is hangared at an airport or if the damage is sustained during the flight to an airport, the best option for repairs will usually be the airport mechanics. If you hangar your aircraft at an airport, you may even receive maintenance and repair discounts or packages that will make it less costly to repair your aircraft.

Have the Plane Ferried or Flown for Repairs

If damage to your aircraft is more extensive or severe than what an airport is able to handle or what a traveling mechanic can fix, it may be necessary to use a ferry or to have it flown on a larger plane to another location. Repairs may require equipment that is not easy to transport to fix a plane. While this may be somewhat inconvenient, it may be the best way to ensure that your plane is safe for flight and that damages have been completely repaired.

Aircraft Maintenance Engineer

Aircraft Maintenance Engineer also known as AME is the professional who works in the engineering and maintenance department of an airline and ensures the safe and airworthy aircraft for the airline they are working for. Aircraft Maintenance Engineer is trained to inspect an aircraft, diagnose the snags and solve it at the earliest. They are the professionals who handle the maintenance of the aircraft including its engines, electrical and navigation systems and certify the plane as airworthy before its take off. Their job starts with a complete routine checkup of the plane to make it land safely at its destination for another checkup.

How to Become an Aircraft Maintenance Engineer

Aircraft Maintenance Engineer Eligibility

Educational Qualication

Candidates who want to pursue Aircraft Maintenance Engineer Course must have completed 10+2 with at least 50% marks in Physics, Chemistry and Mathematics.

Age Limits

The upper age limits are 23 years. However upper age limit is relaxed for those who have done B.Sc Non-Medical after their 10+2 class.

Aircraft Maintenance Engineer Required Skills

- Aircraft Maintenance Engineer should have good mathematical and logical skills.
- They need to be good at organizing, working in teams of cross-sectional skills and as such be able to get on well with people at all levels.
- Applicants should be able to work diligently and patiently so as to do his work with utmost precision as a little mistake can prove fatal for hundreds of lives on board.
- They are also required to constantly update themselves with the latest maintenance techniques and technologies.
- They should also develop good problem-solving skills; have excellent mathematical precision and communication skills.
- They should also be able to apply the principles of science and technology to create aircraft, components and support equipment.
- An Aircraft Maintenance Engineer must be able to resolve issues in quick times into making plane airworthy in time.

How to become an Aircraft Maintenance Engineer

For becoming an Aircraft Maintenance Engineer one has to follow the given steps-

Step 1

After completing their 12 class with Physics, Chemistry and Mathematics as the main subjects, they pass B.E. / B.Tech. in Mechanical or Avionics trades.

They can enrol for a Three Years Aircraft Maintenance Engineering Training Programm having two streams i.e Mechanical – covers jet and piston engines as well as light and heavy aircraft- and Avionics-covers radio navigation, and electrical and instrument system.

The first year of AME training provides basic information about the aircraft rules and regulations.

In the second year, general engineering and maintenance are taught to the students that include subjects like aerodynamics or theory of flight, metallurgy, electronics, handling sophisticated equipment and practical work in machine rooms and on aircraft engines.

In the third year, the training is focused on specific areas like light aircraft, heavy aircraft, piston engine, jet engines or helicopters.

After three years of study, candidates are awarded Basic Aircraft Maintenance Training Certicate followed by an examination by the Director General of Civil Aviation. The successful candidates have issued a license to service the aircraft.

Step 2

After getting the DGCA-AME license the candidates are required to obtain a regulatory license or approval on a particular type of aircraft. If it is an Airbus or Boing one has to obtain an individual license to service that particular type of aircraft.

Step 3

Students of AME course can also appear and qualify as AMAeSI. It is considered as equivalent to B.E or B.Tech. program in Aeronautical Engineering and candidate can go for M.Tech.

Aircraft Maintenance Engineer Job Description

- The main objective of an Aircraft Maintenance Engineer's job is the safe journey of an Aircraft from its take off to its landing. For this, every Aircraft need meticulous and consistent maintenance and grooming to full this objective.
- An Aircraft Maintenance Engineer handles the maintenance of an aircraft that includes its engine as well as its electrical and navigation systems. They are involved in overhauling of grounded aircraft at airports, hangers and workshop. Their job profile includes replacing the plane parts, treating corrosion and checking tyres, instruments, wheel aps etc. They also perform certain diagnostic and mechanical duties covering repair, maintenance, overhaul, troubleshooting besides performing inspections and modification on the aircraft.
- They got opportunities to perform all these duties on sophisticated aircraft systems including commercial jets, military fighter planes, Helicopters and another type of flying machines.
- To do their work flawlessly they are well trained to use specialized aviation and power tools, computers, diagnostic equipment and x-ray machines. While performing their duty they may climb steep ladders or stand on scaffolds, manage other maintenance staff and spend long hours on emergency repairs.
- Aircraft Maintenance Engineer is also authorized to inspect the aircraft thus they also perform the required checks and certify a plane fitness

Aircraft Maintenance and Repair

The successful utilization of new materials and structural concepts relies on maintenance programs that cost-effectively ensure passenger safety.

Airline Maintenance

Maintenance programs are evolved and developed for each new type of aircraft based on previous experience with similar materials, engines, components, or structures. New materials or structures, for which experience is limited, are observed more frequently until a basic level of confidence is established. Time extensions to inspection intervals are based on observations made during routine service checks. A typical airline maintenance and service plan is outlined in table. The objectives of an effective maintenance program are as follows:

- Ensure, through maintenance activity, that the inherent safety and reliability imparted to an aircraft by its design are sustained.
- Provide opportunities to restore levels of safety and reliability when deterioration occurs.
- Obtain information for design modification when inherent reliability is not adequate.
- Accomplish the above at the lowest possible cost.

Structural Maintenance

Any new aircraft program is based on assessing structural design information, fatigue and damage tolerance evaluations, service experience with similar aircraft structures, and pertinent test results. Generally, the maintenance task evaluates sources of structural deterioration including accidental damage, environmental deterioration, and fatigue damage; susceptibility of the structure to each source of deterioration; the consequences of structural deterioration to continuing airworthiness including effect on aircraft (e.g., loss of function and reduction of residual strength, multiple-site or multiple-element fatigue damage, the effect on aircraft flight or response characteristics caused by the interaction of structural damage or failure with systems or power plant items, or in-flight loss of structural items); and the applicability and effectiveness of various methods of detecting structural deterioration, taking into account inspection thresholds and repeat intervals.

Component Maintenance

The application of new materials will not cause undue maintenance difficulties or hardship for the airlines provided the aircraft designer is familiar with component experience. Airline experience indicates that hardware items wear out, but statistical old-age wear-out in complex mechanical, electrical, and avionic components is not a dominant pattern of failure. In fact, over 90 percent of generic part types show either random distribution of failure or gradually increasing probability of failure with age.

The reliability of a part or component of aircraft hardware is only as good as its inherent design (supported by adequate maintenance) allows it to be. Hence, it is generally accepted that (1) good maintenance allows parts to reach their potential reliability; (2) overmaintaining does not improve reliability, but does waste money; and (3) undermaintaining can degrade reliability. In general, fundamental design changes are required to correct inherent component reliability problems.

There are three approaches to preventative maintenance that have proven to be effective. The first method, hard time, involves removing a unit from service when it reaches a pre-ordained parameter value. The second method, functional check or inspection, involves monitoring a characteristic dimension or usage/operating parameter of a piece of hardware to determine if it is still suitable for continued operation, or if it should be removed to prevent an in-service failure. The third method, functional verification, requires performing an operational check of hardware function(s) to determine each function's availability if it is normally hidden from the scrutiny of the flight and operating crew.

There are many components for which measurement of deterioration, periodic removal for maintenance, and hidden function verification are not economically feasible or beneficial. Such parts require routine performance or reliability monitoring, and no preventive maintenance is required or desirable. Modern aircraft are more tolerant of failures than older aircraft designs because of the increased redundancy provided in the design.

Table: Typical Airline Maintenance and Service Plan

When Service is Performed	Type of Service Performed	Impact on Airline Service
Prior to each flight	"Walk-around"—visual check of aircraft exterior and engines for damage, leakage, and brake and tire wear	None
Every 45 hours (domestic) or 65 hours (international) flight time	Specific checks on engine oils, hydraulics, oxygen, and specified unique aircraft requirements	Overnight layover service
Every 200–450 hours (22–37 days) flight time	"A" check—detailed check of aircraft and engine interior, services and lubrication of systems such as ignition, generators, cabin, air conditioning, hydraulics, structure, and landing gear	Overnight layover service
Every 400–900 hours (45–75 days) flight time	"B" check (or "L" check)—torque tests, internal checks, and flight controls	Overnight layover service
Every 13–15 months	"C" check—detailed inspection and repair program on aircraft engines and systems	Out of service for 3–5 days
Every 2 years (narrow-body aircraft)	Inspection and reapplication of corrosion protective coatings	Out of service up to 30 days
Every 3–5 years	Major structural inspections with attention to fatigue damage, corrosion, etc. Aircraft is dismantled, repaired, and rebuilt. Aircraft is repainted as needed	Out of service up to 30 days

Generally, most airlines classify specific component maintenance tasks as follows:

- Lubrication or servicing, where the replenishment of the consumable reduces the rate of functional deterioration;
- Operational or visual check, where identification of the failure must be possible;
- Inspection or function check, where reduced resistance to failure must be detectable and the rate of reduction in failure resistance must be predictable;
- Restoration, where the item must show functional degradation characteristics at an identifiable age, have a large proportion of units survive to that age, and be able to be restored to a specific standard of failure resistance; and
- Discard, where the item must show functional degradation characteristics at an identifiable age, and a large proportion of units are expected to survive to that age.

Malfunctions of components should be evident to the operating crew, have no direct adverse effect on safety (whether they occur as a single or multiple event), and minimize the effect on the operation of the aircraft itself.

Service Experience

Effective application of new materials on commercial aircraft requires the designer to consider potential sources of damage or degradation in operating environments and to develop a maintenance and repair approach to address them. Damage may occur due to flight loads, thermal and environmental cycles, and aircraft operation and servicing activities. A number of valuable lessons have been learned from previous experience with metallic and composite structure in the current fleet.

Ramp and Maintenance Damage

An International Air Transport Association survey estimates that 36–40 percent of damage to aircraft is from ramp and maintenance damage, sometimes called friendly foreign object damage. Figure shows a diagram of the Boeing 777 aircraft interfaces with servicing and other equipment. These areas are especially prone to damage and require robust material performance in these locations.

To determine the extent of groundhandling damage, 11 airline operators were queried for ground damage history during the years 1990 to 1993. Of the 2,241 incidents reported, more than a third were from unknown causes. A tabulation of the causes of damage is given in table.

Ramp and maintenance damage can represent significant costs to the airlines. The repair of a damaged component is only part of the cost. The airline also bears the cost of flight delay or cancellation and the effects on connections and aircraft rotations.

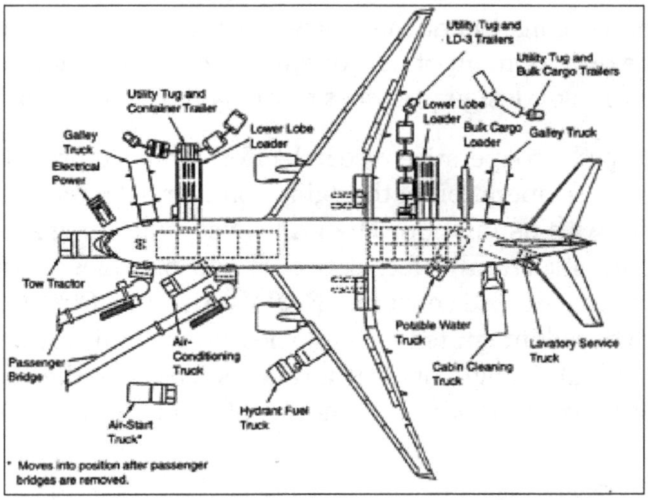

Diagram of aircraft interfaces with servicing and other equipment.

Table: Causes of Ground Damage to Aircraft

Cause of Failure	Number of Incidences
Unknown	773
Catering	137
Belt loader	122
Loader	101
Lavatory and water service	66
Container	44
Jetway	246
Baggage cart	127

Tug/towbar/taxi	109
Maintenance	86
Cargo loading	50
Fueling	33

Aging Aircraft

In April 1988, an Aloha Airlines Boeing 737-200 experienced an in-flight structural failure in which a large section of the upper fuselage ripped open and separated from the aircraft. The failure resulted from multiple-site damage (MSD) and corrosion. In this case, MSD was the link-up of small fatigue cracks extending from adjacent rivet holes in a longitudinal lap joint in the fuselage. The accident focused international attention on the problems of operating an aging commercial fleet.

In 1990 approximately 46 percent of the U.S. commercial air transport fleet was over 15 years old, and 26 percent was over 20 years old. If current usage and replacement trends continue, the number of aircraft over 20 years old will double by the year 2000. Currently some 3,200 aircraft are affected by FAA Airworthiness Directives that concern operation and maintenance of the aging fleet. The review of experience with aging aircraft has caused an increase in the emphasis on stress corrosion, corrosion, fatigue, and MSD issues. This experience has caused, in turn, the selection of new aircraft alloys with better constituent chemistry control or changes in heat treatment tempers. Also, it has stimulated the development of new organic finishes that significantly retard corrosion, as well as the implementation of design practices to vastly improve corrosion resistance.

The FAA and NASA developed a cooperative research effort aimed at providing a technological basis for ensuring the continued safe operation of the aging commercial aircraft fleet. Each agency developed a program consistent with its mission. The FAA's National Aging Aircraft Research Program addresses the aging aircraft structural safety concerns and provides certification authorities and operators with the tools to meet those concerns. NASA's Airframe Structural Integrity Program is focused on developing advanced integrated technologies to economically inspect for damage and to analytically predict the residual strength of older airplanes. Together these programs form the technological basis for a cooperative effort with U.S. industry to address the critical aging aircraft issues.

Multiple-Site Damage

MSD is a form of widespread fatigue damage that is characterized by small cracks emanating from structural details such as fastener holes. If cracks emanate from adjacent fastener holes, they have the potential to link up and lead to unexpected catastrophic failures. Also, even without link-up, multiple-site cracks can severely degrade the capability of the structure to withstand major damage from other discrete sources.

In the past, the standard industry practice was to visually inspect the airframe for damage. Various levels of inspections ranging from daily walk-around inspections to detailed tear-down inspections were performed. Instrumented nondestructive evaluation (NDE) methods such as eddy current probes were used only to inspect local regions of the structure where previous cracking problems had occurred. While these inspection methods were labor intensive and highly subjective, they were acceptable because the airframe was designed to survive a two-bay skin crack with a severed frame or stiffener. This design criterion was established to enable the airplane to tolerate major discrete source damage (i.e., such as might be encountered as a result of an engine structural failure) as well as large cracks resulting from the link up of smaller fatigue cracks or the unstable propagation of manufacturing flaws or other service-induced damage. Such damage is large enough that it should be easily detected, and the operator does not need to search for small cracks to ensure the structural integrity of the airframe. However, this "fail-safe" philosophy assumed that the structure adjacent to the major damage (e.g., the two-bay crack) was free of MSD. Design residual strength requirements were based on this assumption. However, the existence of very small cracks (e.g., a few hundredths of an inch or tenths of a millimeter in length) in the adjacent structure can severely degrade this residual strength and thus jeopardize the safety of the airplane as it did in the Aloha Airlines incident. Therefore, inspection of aging aircraft has become much more onerous than for newer aircraft because safety is vitally dependent on the detection of the very small cracks associated with this onset of MSD. This represents a major challenge to the inspection and aircraft industries.

The principal technical needs are (1) to develop and verify advanced NDE technology that can reliably and economically detect disbonds, small MSD fatigue cracks, and corrosion and characterize their effect on the residual strength; and (2) to develop and verify advanced fracture mechanics and structural analysis methodology to predict fatigue crack growth and residual strength of airframe structures to determine in-service inspection thresholds and repeat intervals, quantitatively evaluate inspection findings, and design and certify structural repairs.

Corrosion

Corrosion of aging aircraft has been described as an insidious problem (Marceau, 1989). While other aging mechanisms, such as wear and fatigue, are somewhat predictable and can be addressed by the airline maintenance programs to preclude major structural problems, corrosion—especially in its localized forms—is very difficult to predict and detect. Factors that influence the extent of corrosion on aircraft are materials selection, design, component processing and finishing, operational environments, and maintenance programs.

It is anticipated that airplanes manufactured today will experience fewer corrosion problems than those in the current aged fleet because of significant design and corrosion protection improvements that have been implemented and because of operators' increased awareness of the role of these improvements in preventive maintenance. Clearly, maintenance and corrosion control programs will continue to play a major role in the control of corrosion as airplanes age.

Some examples of design improvements to reduce corrosion on the Boeing 777 include:

- Enhanced drainage, especially in the keel of the aircraft;
- The sealing of faying surfaces in corrosion-prone areas;
- The application of improved finish systems;
- Liberal use of corrosion-preventive compounds;
- Implementation of a good corrosion control maintenance program; and
- Improved access for inspection of corrosion-prone areas.

Major airline fleets include aircraft ranging in age from new to 25 years old. Consequently, the degree of corrosion protection incorporated into the airplane varies from limited protection for older aircraft to fairly extensive protection for newer aircraft. Corrosion control programs are tailored to individual fleets, depending on age, prior experience, flight environment and degrees of corrosion protection incorporated prior to the delivery of the aircraft. All protective finishes are maintained and corrosion prevention compounds are applied during periodic maintenance. Critical areas that are prone to excessive corrosion include areas below the galleys, doorways, lavatories, cargo compartment subfloors, inside external fairings, and the bilges which are all treated at four-year intervals. Landing gear wheel wells and wing spars are treated yearly. Longer intervals of time are allowed between reapplications of corrosion prevention compounds in the case of less-severe environments.

Aging aircraft repairs have typically involved upper-skin lap fastener replacement, nonbonded skin panel replacement, skin lap doubler repairs, frame reinforcement, entryway door and scuff-plate doublers, replacement bushings and clevis joints, bulkhead forging replacement, and selected landing gear component replacement. Based on service experience, the airlines have expectations that manufacturers of new aircraft will:

- Include stress corrosion prevention in all design reviews with airline customers,
- Assemble all structure below the aircraft floor with sealant on all faying surfaces,
- Coat all detail parts with corrosion-inhibiting primer or polyurethane topcoat before assembly,
- Not use adhesive-bonded fuselage skin panels below the floor line or in areas subject to severe corrosion environments such as galleys and lavatories,
- Treat all basic fuselage structure with corrosion-preventive compounds, and
- Perform a complete (100 percent) inspection for delamination of bonded skin panels prior to the aircraft delivery in order to establish a baseline for subsequent inspection.

The objective of aging aircraft programs is to ensure the continued airworthiness of large transport aircraft as long as they remain in commercial service. Because new materials and fabrication processes may yield different degradation and damage mechanisms, a preproduction review should ensure that the new aircraft design includes lessons learned from the existing aging fleet.

Many of the steps needed to improve aging performance are detailed below. Most of these steps have now been incorporated into recent aircraft designs. The susceptibility of aircraft to corrosion and MSD fatigue can be reduced by the following steps:

- Eliminating cold-bond lap-joint design details;
- Providing adequate drainage to eliminate corrosion in areas where moisture accumulates;
- Using the most corrosion-resistant materials and tempers available;
- Evaluating galvanic couples with typical coating damage;

- Testing dissimilar materials design details for size effects in areas with joints;
- Controlling design stress levels, improving design details, and utilizing improved manufacturing and maintenance procedures to preclude the onset of multiple-site damage within the operational lifetime of the airplane;
- Developing predictive and monitoring techniques for the onset of MSD (simple and cost-effective techniques should be integrated into the aircraft maintenance plan); and
- Providing a complete corrosion prevention and control program within the aircraft maintenance program upon delivery of an aircraft.

The present focus on aging aircraft will lead to better corrosion-resistant treatments for next-generation aircraft. Materials selection in wet areas, the design drainage schemes, the use of insulation standoffs, and sealing and finishing systems have all been improved. The benefits of these improvements should be evident during in-service performance of the Boeing 777 and future aircraft. Liberal use of corrosion-preventive compounds applied in the aircraft assembly process and periodically in service, using a good corrosion control maintenance program, should minimize future corrosion concerns.

Structural Composites

As discussed in chapter 4, prior to the latest generation of aircraft, which includes the Airbus A320 and the Boeing 777, structural composites have been used on aircraft flight control surfaces such as elevators, spoilers, ailerons, and rudders, as well as fairing and fillet panels, landing gear doors, engine cowl doors, and other secondary structures. For these applications, honeycomb sandwich designs with thin 0.6—1.5 mm (0.024—0.060 in.) composite facesheets are most common. It follows that most of the experience with advanced composites has been obtained with this kind of construction. Previously, similar constructions with fiberglass skins and nonmetallic honeycomb core have been used. There is much less service experience with thicker-skin laminate designs that have been used in composite primary structure.

In general, the service experience with composites indicates that damage occurs because of discrete sources such as impacts, lightning strikes, and handling rather than progressive growth caused by a fatigue condition. In addition to groundhandling damage, a recent survey by the International Air Transport Association, summarized in table, lists the particular causes of damage that occur in the current generations of composite structure.

The types of damage to composite components include disbonds or delaminations (45 percent), holes or punctures (35 percent), cracks (10 percent), and other damage (10 percent). An especially difficult maintenance issue resulting from these types of damage is when perforation allows the incursion of hydraulic fluids, water, and other liquids into the honeycomb core. Composites may also suffer loss of load-bearing capability due to resin charring and the potential for corrosion of adjacent metallic surfaces. Typical causes of composite service damage mechanisms are shown in table.

Service experience with thicker composite laminate constructions, such as that used on primary structures on the Airbus A320 and Boeing 777, is not adequate enough to establish damage trends.

Composite Repair

The current methods used by the airlines to repair damage to aircraft composite structure (secondary structure and primary flight controls) depend on the extent of damage, the time available to perform the repair, and the time until the next scheduled maintenance visit. In approximately 80 percent of all cases, the damage is covered with adhesive-backed aluminum foil ("speed tape") or temporarily repaired and deferred for a specific time to provide for interim or permanent repair or part replacement. Occasionally, temporary or permanent repairs can be performed by bonding or bolting a sealantcoated metal or precured composite overlay over the damage. Finally, most permanent repairs are accomplished with room-temperature curing, wet lay-up and precured patch techniques. Other permanent repairs use prepreg that cures under vacuum or autoclave pressures at temperatures lower than the cure temperature of the original structure. Repair resins are being developed that have relatively low cure temperatures, but have thermal and environmental resistance similar to higher-temperature curing systems.

Tabl:e Most Common Causes of Composite Structure Damage to Aircraft

Cause of Failure	Percent of Incidences
Moisture and chemical fluids attack	30
Other (heat damage, fatigue, abrasion, and erosion)	11
Bird strikes and hail damage	8
Runway rocks and foreign object damage	8
Lightning strikes	7

The thicker laminate construction used in composite primary structure are not conducive to wet lay-up patch technologies. Thin facesheets on honeycomb panels are currently repaired using bonded scarf patches with a scarf taper of 20:1. For thicker constructions the result would be the removal of a large amount of undamaged material. The emphasis in the development of primary structure repairs has therefore been on fastened, precured composite or metallic splice plates, similar to current metal repair techniques. A design for a fairly complex bolted repair is shown in figure. The issues that must be addressed in these types of repairs include (1) criteria for determining when repairs are required; (2) availability of standardized repair elements; (3) drilled hole quality; (4) ability to restore original strength, durability, and damage tolerance; and (4) ability to match existing contours.

Given typical flight schedules, composite structure repair must be accomplished within eight hours on an overnight layover, otherwise the part would be replaced with a spare. Repairs carried out during an overnight stop (at line stations or hubs) and repairs requiring more-intensive maintenance center rework should follow guidelines established by the manufacturer's structural repair manual and the appropriate industry group, the Commercial Aircraft Composite Repair Committee. Because composite structures are fabricated from a large number of resin/fabric systems from several qualified suppliers worldwide, it is difficult and expensive for airlines to stock a wide variety of repair materials. Accordingly, there is a pressing need for standardization of repair materials and processes.

While next-generation aircraft are expected to use laminated or tailored composite skin structure that is more damage tolerant, there are many lessons that have been learned in maintaining and repairing the thin-skin, nonmetallic honeycomb sandwich constructions on today's aircraft that need to be considered in new aircraft design and materials selection. Maintainable designs need to consider component accessibility, permitted defect levels, and nondestructive testing techniques. Nondestructive test indications need to be correlated with structural criteria throughout the life of the aircraft.

Table: Causes of Service Damage to Composite Structure

Damage Mechanism	Affected Components
Mechanical Damage (in flight)	
Hail impact	Radome Engine inlet Upper wing and tail plane fixed panels Flight controls
Bird strike	Engine inlet cowl Radome
Engine disintegration	Engine cowl Fuselage Lower wing and tail plane fixed panels
Tire protector separation	Flaps Lower wing-to-body fairings Landing gear doors
Mechanical Damage (on ground)	
Hail impact	All horizontal surfaces (wing panels, flight controls, upper areas of engine cowl)
Groundhandling equipment	Engine cowl Wing and tail leading and trailing edges Landing gear doors
Mishandling	Engine cowl Access doors
Overload due to actuation system failure	Flight controls Spoilers Thrust reverser
Transport and handling	All removable components
Lightning strike	Radome Leading and trailing edge components (aileron, rudder, elevator, leading-edge fairings) Engine cowl
Overheat	Engine cowl Landing gear door (in case of brake overheat)

Erosion	Radome Engine inlet cowl Leading-edge fairings
Chemical contamination	
Skydrol (hydraulic fluid) leakage	Engine cowl Actuated components (flight controls, spoilers)
Paint stripper	All painted components
Corrosion	All aluminum honeycomb with composite faceskins Improperly isolated aluminum brackets, hinges, etc.

Composites must be protected by finishes with resistance to fluid penetration and ultraviolet degradation. significant costs are incurred by the airlines over the life of an aircraft in the maintenance of protective finishes. The durability of protective finish systems (including aerodynamic surfaces) should be characterized prior to production.

The removal of finishes from composites is a slow and expensive process. Since chemical strippers attack the polymer matrix, airlines generally remove finishes through mechanical abrasion processes. New paint removal processes like laser, heat, frozen carbon dioxide blasting, and wheat starch blasting are being evaluated. Rapid, low-cost, on-aircraft paint removal techniques require implementation if larger areas of composite surfaces are to be accepted on the next-generation aircraft.

Fast, durable, temporary, and permanent field and shop repair procedures should be developed for new composites. This would include development of repair materials, tooling, and processes for high-modulus, high-strength composite skins, metallic and nonmetallic honeycomb sandwich structure, and laminated hybrid and bonded metal structures. These repairs should be accomplished without major disassembly of structure—preferably while working from an exposed exterior or interior surface. It is critical to design for accessibility and interchangeability of parts, especially those for parts that have high damage probability.

Types of Aircraft Maintenance

Light or Line Maintenance

This would typically include Pre-flight checks, Daily checks (before first flight) Fluids, Failure rectification as well as Minor, scheduled maintenance tasks as follows. According to EASA Part 145, AMC 145.A.10, Line Maintenance should be understood as "any maintenance that is carried out before flight to ensure that the aircraft is fit for the intended flight." This may include:

- Trouble shooting
- Defect rectification
- Component replacement, up to and including engines and propellers, with use of external test equipment if required
- Scheduled maintenance and/or checks including visual inspections that will detect obvi-

ous failures but do not require extensive in depth inspection. It may also include internal structure, systems and powerplant items which are visible through quick opening access panels/doors

- Minor repairs and modifications which do not require extensive disassembly and can be accomplished by simple means.

EASA Part 145, AMC 145.A.10 also explains that "for temporary or occasional cases (ADs, SBs) the Quality Manager may accept base maintenance tasks to be performed by a line maintenance organisation provided all requirements are fulfilled as defined by the competent authority". It is also noted that "Maintenance tasks falling outside these criteria are considered to be Base Maintenance".

Base or Heavy Maintenance

Base Maintenance may be referred to as Heavy (or Depth) Maintenance, and consists of tasks that are generally more in-depth and long-lasting than those above, but are performed less frequently. An MRO company will have to have large facilities and specialised equipment and staff to undertake base maintenance, and many operators contract-out this function. The different activities may include:

- C and D Checks (Block checks see Maintenance Programme) which will check for deterioration of the airframe, engines and systems, e.g. corrosion, fatigue
- Removal of defects – implementation of Service Bulletins (SB) and Airworthiness Directives (AD), although this can also be done during Line maintenance.
- Technology upgrade – fitting of Terrain Avoidance and Warning System (TAWS), Airborne Collision Avoidance System (ACAS) etc.
- Cabin reconfiguration, painting etc.

Shop or Component Maintenance

The third form of maintenance can be termed as "Workshop" or just Shop maintenance. This covers maintenance on components when removed from aircraft e.g. engines, APU, seats. Sometimes this is carried out within the same organisation as the Base Maintenance, but sometimes special companies carry out this work separately.

Maintenance Intervals

The intervals of maintenance are parameters set within the Approved Maintenance Schedule (AMS), which is in turn based on the Maintenance Planning Document (MPD). These will be set according to different criteria, mostly depending on how well damage can be detected and failure predicted.

Hard Time

- "Preventative process in which known deterioration of an Item is limited to an acceptable level by the maintenance actions

- Carried out at periods related to time in service (e.g. calendar time, number of cycles, number of landings).

On-condition

- "Preventative process in which Item are inspected or tested, at specified periods, to an appropriate standard to determine whether it can continue in service
- Such an inspection / test may reveal a need for maintenance action.
- Fundamental purpose of On-Condition is to remove an Item before its failure in service."

Condition Monitoring

- "Information on Items gained from monitoring is collected, analysed and interpreted on a continuing basis as a means deciding whether or not to implement corrective procedures."
- This process is normally automated and may form part of the aircraft's on-board health management system.

Units for Maintenance Intervals

- Flight Hours (FH), for items that are in constant operation e.g. Fuel Pumps, Electric Generators
- Flight Cycles (FC), for items operated once or twice per flight e.g. Landing gear, air starter, brakes, hull pressurisations
- Calendar Time (Cal), for items exposed whether operated or not e.g. Fire Extinguishers, Corrosion to Structure
- Operating hours, for items not operated every flight, or otherwise independent of FH or FC e.g. APU operation.

How to Maintain an Aircraft

Maintenance of an aircraft is of the utmost importance for safe flight. Licensed technicians are committed to perform timely maintenance functions in accordance with the manufacturer's instructions and under the 14 CFR. At no time is an act of aircraft maintenance taken lightly or improvised. The consequences of such action could be fatal and the technician could lose his or her license and face criminal charges.

Airframe, engine, and aircraft component manufacturers are responsible for documenting the maintenance procedures that guide managers and technicians on when and how to perform maintenance on their products. A small aircraft may only require a few manuals, including the aircraft maintenance manual. This volume usually contains the most frequently used information required to maintain the aircraft properly. The Type Certificate Data Sheet (TCDS) for an aircraft also contains critical information. Complex and large aircraft require several manuals to convey correct maintenance procedures adequately. In addition to the maintenance manual, manufacturers may produce such volumes as structural repair manuals, overhaul manuals, wiring diagram manuals, component manuals, and more.

Note that the use of the word "manual" is meant to include electronic as well as printed information. Also, proper maintenance extends to the use of designated tools and fixtures called out in the manufacturer's maintenance documents. In the past, not using the proper tooling has caused damage to critical components, which subsequently failed and led to aircraft crashes and the loss of human life. The technician is responsible for sourcing the correct information, procedures, and tools needed to perform airworthy maintenance or repairs.

Standard aircraft maintenance procedures do exist and can be used by the technician when performing maintenance or a repair. These are found in the Federal Aviation Administration (FAA) approved advisory circulars (AC) 43.13-2 and AC 43.13-1. If not addressed by the manufacturer's literature, the technician may use the procedures outlined in these manuals to complete the work in an acceptable manner. These procedures are not specific to any aircraft or component and typically cover methods used during maintenance of all aircraft. Note that the manufacturer's instructions supersede the general procedures found in AC 43.13-2 and AC 43.13-1.

All maintenance related actions on an aircraft or component are required to be documented by the performing technician in the aircraft or component logbook. Light aircraft may have only one logbook for all work performed. Some aircraft may have a separate engine logbook for any work performed on the engine(s). Other aircraft have separate propeller logbooks.

Large aircraft require volumes of maintenance documentation comprised of thousands of procedures performed by hundreds of technicians. Electronic dispatch and recordkeeping of maintenance performed on large aircraft such as airliners is common. The importance of correct maintenance recordkeeping should not be overlooked.

Location Numbering Systems

Even on small, light aircraft, a method of precisely locating each structural component is required. Various numbering systems are used to facilitate the location of specific wing frames, fuselage bulkheads, or any other structural members on an aircraft. Most manufacturers use some system of station marking. For example, the nose of the aircraft may be designated "zero station," and all other stations are located at measured distances in inches behind the zero station. Thus, when a blueprint reads "fuselage frame station 137," that particular frame station can be located 137 inches behind the nose of the aircraft.

To locate structures to the right or left of the center line of an aircraft, a similar method is employed. Many manufacturers consider the center line of the aircraft to be a zero station from which measurements can be taken to the right or left to locate an airframe member. This is often used on the horizontal stabilizer and wings.

The applicable manufacturer's numbering system and abbreviated designations or symbols should always be reviewed before attempting to locate a structural member. They are not always the same. The following list includes location designations typical of those used by many manufacturers.

- Fuselage stations (Fus. Sta. or FS) are numbered in inches from a reference or zero point known as the reference datum. The reference datum is an imaginary vertical plane at or near the nose of the aircraft from which all fore and aft distances are measured. The distance to a given point is measured in inches parallel to a center line extending through the aircraft from the nose through the center of the tail cone. Some manufacturers may call the fuselage station a body station, abbreviated BS.
- Buttock line or butt line (BL) is a vertical reference plane down the center of the aircraft from which measurements left or right can be made.
- Water line (WL) is the measurement of height in inches perpendicular from a horizontal plane usually located at the ground, cabin floor, or some other easily referenced location.
- Aileron station (AS) is measured outboard from, and parallel to, the inboard edge of the aileron, perpendicular to the rear beam of the wing.
- Flap station (KS) is measured perpendicular to the rear beam of the wing and parallel to, and outboard from, the inboard edge of the flap.
- Nacelle station (NC or Nac. Sta.) is measured either forward of or behind the front spar of the wing and perpend
- icular to a designated water line.

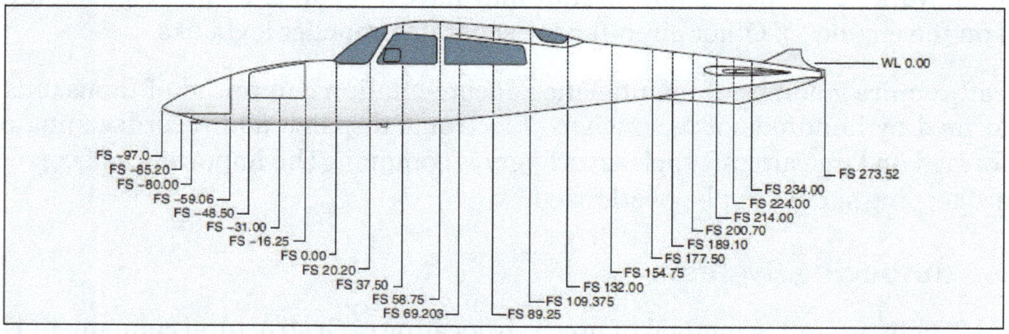

Figure: The various body stations relative to a single point of origin illustrated in inches or some other measurement (if of foreign development).

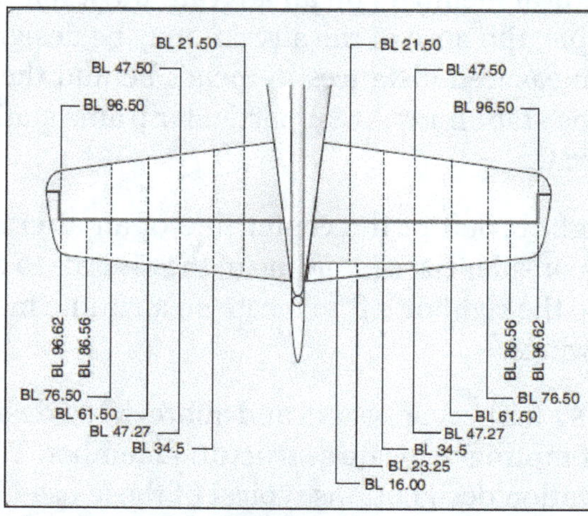

Figure: Butt line diagram of a horizontal stabilizer.

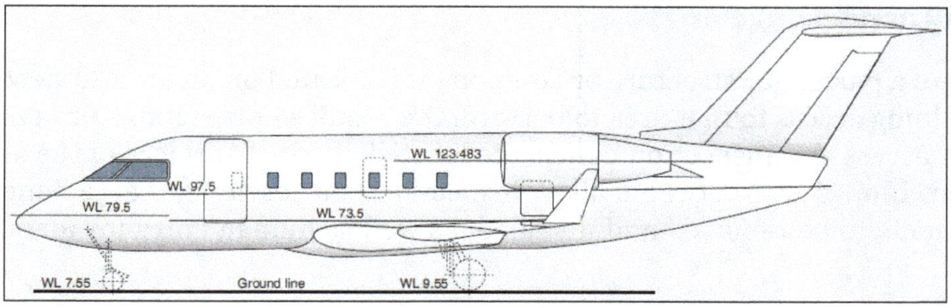

Figure: Water line diagram.

In addition to the location stations listed above, other measurements are used, especially on large aircraft. Thus, there may be horizontal stabilizer stations (HSS), vertical stabilizer stations (VSS) or powerplant stations (PPS). In every case, the manufacturer's terminology and station location system should be consulted before locating a point on a particular aircraft.

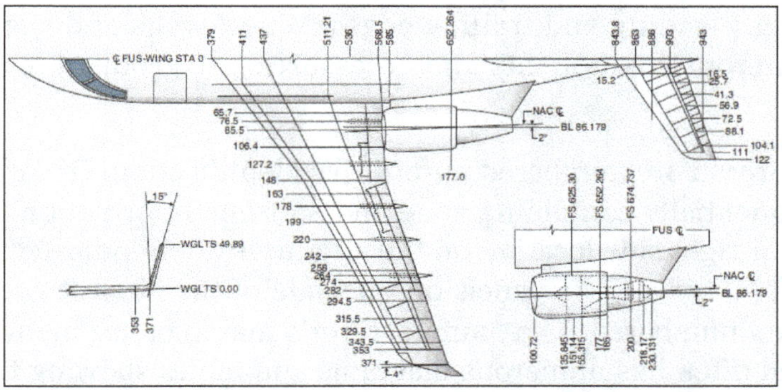

Figure: Wing stations are often referenced off the butt line, which bisects the center of the fuselage longitudinally. Horizontal stabilizer stations referenced to the butt line and engine nacelle stations are also shown.

Another method is used to facilitate the location of aircraft components on air transport aircraft. This involves dividing the aircraft into zones. These large areas or major zones are further divided into sequentially numbered zones and subzones. The digits of the zone number are reserved and indexed to indicate the location and type of system of which the component is a part. Figure illustrates these zones and subzones on a transport category aircraft.

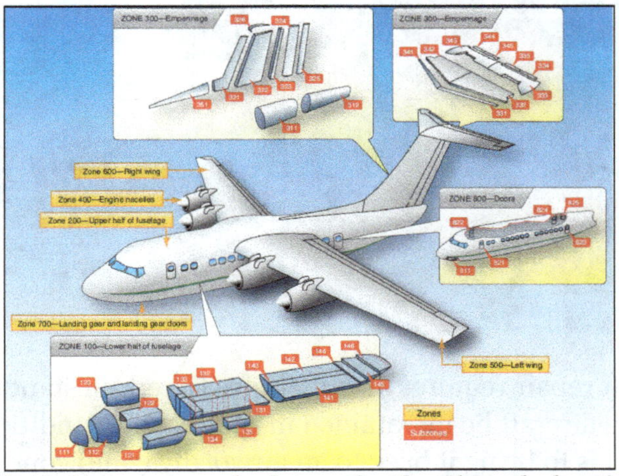

Figure: Large aircraft are divided into zones and subzones for identifying the location of various components.

Access and Inspection Panels

Knowing where a particular structure or component is located on an aircraft needs to be combined with gaining access to that area to perform the required inspections or maintenance. To facilitate this, access and inspection panels are located on most surfaces of the aircraft. Small panels that are hinged or removable allow inspection and servicing. Large panels and doors allow components to be removed and installed, as well as human entry for maintenance purposes.

The underside of a wing, for example, sometimes contains dozens of small panels through which control cable components can be monitored and fittings greased. Various drains and jack points may also be on the underside of the wing. The upper surface of the wings typically have fewer access panels because a smooth surface promotes better laminar airflow, which causes lift. On large aircraft, walkways are sometimes designated on the wing upper surface to permit safe navigation by mechanics and inspectors to critical structures and components located along the wing's leading and trailing edges. Wheel wells and special component bays are places where numerous components and accessories are grouped together for easy maintenance access.

Panels and doors on aircraft are numbered for positive identification. On large aircraft, panels are usually numbered sequentially containing zone and subzone information in the panel number. Designation for a left or right side location on the aircraft is often indicated in the panel number. This could be with an "L" or "R," or panels on one side of the aircraft could be odd numbered and the other side even numbered. The manufacturer's maintenance manual explains the panel numbering system and often has numerous diagrams and tables showing the location of various components and under which panel they may be found. Each manufacturer is entitled to develop its own panel numbering system.

How to Repair an Aircraft

Making any type of aircraft repair requires an adherence to certain standards and procedures. The regulations require that an aircraft be maintained in a airworthy condition. So, when an aircraft is repaired, the ultimate goal is to bring it back to an airworthy (safe) condition.

The FAA classifies repairs into two categories. Major and minor. Following is the FAA's definition of both:

Major repair means a repair:

1. That, if improperly done, might appreciably affect weight, balance, structural strength, performance, powerplant operation, flight characteristics, or other qualities affecting airworthiness; or
2. That is not done according to accepted practices or cannot be done by elementary operations.

Minor repair means a repair other than a major repair.

These definitions often lead to controversy when deciding how to classify a proposed repair. The FAA published an advisory circular AC 120-77 - Maintenance and Alteration Data. In that document is a decision chart that can be used to determine major and minor repairs. That decision chart is shown below:

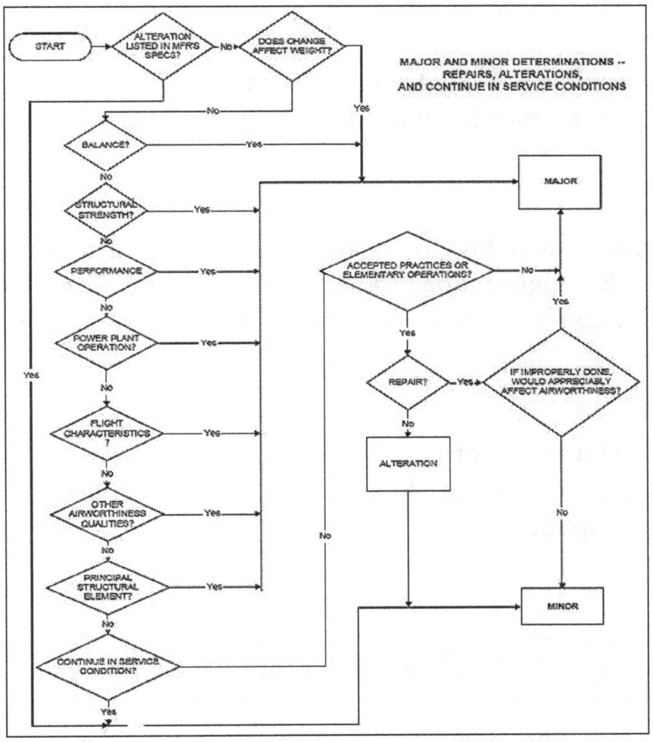

Aircraft Repair and Alteration Decision Chart

Please note the final decision question on the right. "If improperly done would appreciably affect airworthiness?" This is where I believe a minor aircraft repair is sometimes improperly classified.

If you follow this chart and give that last answer some serious thought, you will make the right determination.

Ssome of the different types of repairs that can be made to aircraft:

Aircraft Structural Repairs

Structural repairs are made to aircraft that have sustained damage to the structure (i.e., external skin, stringers, formers, bulkheads, etc.) There are several ways an aircraft structure can be damaged. It is usually caused by ground service equipment, maintenance stands, bird strikes, or lightning strikes.

Component Repairs

Component repairs range from simple part replacements to an entire overhaul. If a component fails to operate properly, it is removed from the aircraft and replaced by one the does operate properly. The removed component is then repaired. These repairs consist of determining the cause of the failure and replacing those internal parts that failed. The component is disassembled, cleaned inspected, repaired, reassembled, tested, and returned to service. These procedures are done in accordance with the component manufacturers maintenance manuals (CMM).

Systems Repairs

Some repairs to aircraft systems (i.e., hydraulic, pneumatic, fuel, etc.) simply consist of replacing a part or component. The faulty part is removed and replaced with a new part.

Engine Repairs

Some limited repairs can be done while the engine is on the aircraft. Those repairs usually consist of parts replacements. Major engine repairs are normally done in a shop that is qualified to completely disassemble, clean, inspect, repair, reassemble and test the engine.

Avionics Repairs

Avionics repairs are treated a bit differently than other components because of their delicate circuit boards and small fragile electrical or electronic parts. Special clean rooms are used to perform repairs to these types of components.

Importance of Aircraft Maintenance

Aircraft maintenance is one of the most important activities that aircraft owners should never underestimate. This is a set of activities that include inspection, reformation and repair of an aircraft. As the owner of an aircraft, you need to know every detail about maintenance if you want to save money and avoid failure in your aircraft system. Aircraft maintenance is not only about replacing a part that is already damaged; this is also about cleaning and refueling. This should be done not only to large planes but also to smaller aircrafts. Maintaining your aircraft is done as a compliance with the rules on aircraft ownership. This is the reason why no company is exempted when it comes to this. There are different airworthiness authorities in the world whose job is to make sure that every owner is doing his part in maintaining the safety and good condition of his aircraft.

Why Aircraft Maintenance is Important

Here, you will find answers on why aircraft maintenance is highly regulated in every part of the world. Here are the different reasons why this activity is important:

- Helps avoid lost flights because of failure. One of the main steps in completing this activity is the replacement of batteries and other parts that are damaged. This is done because batteries have certain life span. If you want to make sure that everything is in its perfect status, you have to schedule regular aircraft maintenance. The replacement of parts is also done in accordance to the recommendation of the manufacturers.
- To maintain good performance. The reason why there is regular maintenance is to make sure that the aircraft is going to work well every time it flies. One of the main parts of this activity is the airflow check. This is done in order to make sure that there is proper airflow around the whole component. When done, this willensure that the aircraft will perform well. All in all, an operational check is made before every flight in order to make sure that failure will not happen when the aircraft sets off.
- To ensure passenger safety. Another reason for aircraft maintenance is to avoid accidents that can be caused by failures. This is mainly done in order to protect passengers. The main concern of every airline and aircrafts is to deliver the passengers to their destination safely.
- Helps extend the life of the aircraft. And finally, an aircraft needs regular maintenance to prolong its life and to make sure that it will perform well for the expected period of time.

There are different levels of inspection that are done when it comes to an aircraft and its project management. There is routine maintenance that is done regularly and there are also other inspections that are done for a period of two months.

Maintenance Checks and Scheduling Tasks

With the constant need for maintenance, an aircraft company should have enough employees to do the tasks that are necessary for the activity's completion. The main goal of maintenance is to make sure that the aircraft owner is not breaching any airworthiness regulations.

There are two very important employees involved in the completion of maintenance. The company should have both a planner and a scheduler. The latter will be working to make a plan that will be fast and efficient in allowing all the scheduled jobs to be done in a reasonable time frame. The job of the scheduler is not really final. After they create a schedule, planners will audit the plan that they have created. They will start to audit not only the plans but also the possible output. This is the best way to determine if the methods that they will be using are perfect for the whole process.

Scheduling tasks is the end part of planning. Before the workers have the schedule of activities, they first have to check the plane in order to know if the maintenance activities will be successful. The difficulty in the process is just one of the determinants of how important it is for aircraft owners to check and maintain their aircrafts.

How to Prepare your Airplane for Winter

Step 1: Wash, Wax, and Detail the Exterior

This may be your last comfortable opportunity to clean and protect the paint and windows of your aircraft. Cleaning off all of the residual bugs, dirt, and oxidation has benefits beyond the smile it will put on your face every time that you look at your airplane. Snow and ice readily cling to contaminants on your aircraft's exterior. A freshly waxed exterior, in contrast, readily sheds ice and snow at least on the ground during preflight. Don't forget the windows. Using an aircraft-specific window cleaner and polish helps protect and seal those as well.

Step 2: Change the Oil

Changing your engine's oil removes harmful acids and contaminants and, with them, any water that is in the mix (at least until your next flight). The oil is doing its job by collecting these contaminants, but it pays to go into the winter with a clean slate. It will also get this task out of the way early in the season so you don't have to deal with blue fingers while trying to do the job later. If you use straight weight oil, this is your opportunity to switch to a thinner oil for the cold temperatures. Since cold temperatures and winter weather often mean less flying, consider an anti-corrosion additive such as Camguard. I use Camguard regularly and have been very satisfied with the results. Don't forget to do oil sampling and trend analysis with every oil change.

Step 3: Lubricate the Airframe

Most aircraft maintenance manuals include a lubrication schedule and chart. Most aircraft owners have never seen it. Take the time to find yours, obtain the lubricants listed, and apply them according to the instructions. Grease, oil, and other prescribed lubricants are your best protection against excessive wear and corrosion for every mechanism on the aircraft. Lubrication also helps avoid putting unnecessary strain on components that become stiffer to operate in the cold.

Step 4: Charge, Maintain, or Replace Batteries

Batteries don't perform well in low temperatures, so it's important to ensure they are in top condition as the temperatures drop. Beginning with the aircraft's main battery, have it tested to ensure it is still performing at the required capacity. If it fails a capacity check, replace it. Consider a sealed AGM battery such as the ones made by Concorde. They reduce battery maintenance because they don't have to be topped off with water. I also recommend using a high-quality, aircraft-specific trickle charger such as a Battery Minder. And, don't stop at the aircraft's main battery. Replace every other battery you can think of, including headset batteries, flashlight batteries, and carbon monoxide detector batteries.

Step 5: Inspect the Heating System

Chances are you'll be using your aircraft's heating system for the first time in many months. Most single-engine aircraft derive their heat from the exhaust system, so inspect it carefully to ensure that carbon monoxide cannot infiltrate your breathing air in flight. If you have a twin, now is the time to make sure that combustion heater is safe and ready for use.

Step 6: Prepare for Preheating

If you already have an installed engine preheater, test and inspect it to make sure it is working properly. If you don't have one, consider the convenience one of these systems might provide. If you use a portable preheater, get that ready for use as well. Finally, consider an in-cabin heater. I have a small, fire-safe ceramic heater on a special thermostat that I use in the cabin to keep the interior temps above freezing and reduce the expansion/contraction cycling on the interior plastics and avionics.

Be Prepared

Inevitably, we gather all sorts of items in the cabin of our aircraft throughout the year: extra batteries, oil, flashlights, notepads, water bottles, and other extraneous items. This is a great opportunity to take every single portable object out of the aircraft, evaluate it, and re-load the aircraft with the items most important to winter flying. Be sure to include cold-weather survival gear such as an emergency blanket, lighter, and other items. Be prepared for an off-field landing and a potentially cold night in the wilderness.

Preparing for winter is a great excuse for a day at the airport bonding with your airplane. With a little preparation, your winter time aloft will be fun, efficient, and enjoyable.

Airworthiness

Airworthiness has a number of aspects which relate to the legal and physical state of an aircraft. According to the Federal Aviation Administration (FAA), the term Airworthy "is when an aircraft or one of its component parts meets its type design and is in a condition for safe operation."

"Airworthiness is the ability of an aircraft or other airborne equipment or system to be operated in flight and on the ground without significant hazard to aircrew, ground crew, passengers or to third parties; it is a technical attribute of materiel throughout its lifecycle."

The connection between airworthiness and flight safety is an obvious but complex one. The design activity, besides meeting the applicable certification code, often seeks to improve the aircraft's economics and cost benefit to both the manufacturer and the operator. Certification authorities will therefore examine all aspects of the design and construction of an aircraft, even when there is apparent improvement to minimum standards. When an aircraft type is first judged to meet all the certification requirements it will be issued with a Type Certificate (TC).

Deficiencies in airworthiness may be indicated following an in-service incident or accident. These may relate to unknown failures, errors or limitations of the Type design and/or failure to meet the conditions for safe operation. In 2002, a cross-party, international study group analysed a series of accidents in an attempt to learn lessons for the certification process. Known as the Commercial Airplane Certification Process Study, their work is a useful meta-analysis of interfaces between certification, operations and maintenance, and produced 15 findings and two observations for suggested improvements.

Defects, Failures and Threats

A defect may have a significant effect on safety and, if not rectified, or only partly rectified, may also be a cause for an accident at a later time. Inappropriate crew actions in response to a malfunction which arises in flight may also lead to a worse outcome. In such cases, an investigation should analyse the crew responses as well as the underlying airworthiness issues. However, in many cases, the flight-crew successfully recovers an aircraft after anomalies occur. The following causal factors are taken from the UK Military Aviation Authority (MAA) Handbooks for Structural Integrity, Systems Integrity and Propulsion Integrity. This is a long list of possible failures or threats which may affect the airworthiness of an aircraft. Some are physical process e.g. overload and fatigue, whereas some relate to human factors and obsolescence.

- Overload
- Operation outside the certificated limits such as those laid down for flight in ice or snow conditions.
- Component degradation due to fatigue, creep, fretting, wear or corrosion, depending on

the system or component

- Accidental Damage (AD) and Environmental Damage (ED)
- Procedural (Design, Manufacturing, Maintenance or Supply) error and Human Factors
- Inadequate or incomplete maintenance.
- Errors in maintenance which may result in a fault becoming obvious a long period after the error was originally made.
- Ageing components
- Change of usage or unmonitored operation
- Lack of configuration control
- Obsolescence and/or legislation change
- Fuel and fuel system hazard
- Lack of adequate oversight of the operator, its practices and policies including training, operation and maintenance by the regulator.
- Deficiencies in the process which led to the issue of the original aircraft Type Certificate.

Defences

The first defence is the process of aircraft type certification, leading to the issue of the Type Certificate. This work is documented so that it remains an accessible foundation for the continuing airworthiness of the aircraft type thereafter. Wherever practicable, the original design will embody redundancy features; i.e. an allowance for the failure of a system or component without any reduction in airworthiness. In some cases, the failure only becomes observable after an aircraft has landed, and requires rectification before further flight.

In more extreme cases a major failure, such as an in-flight failure of an engine on a multi-engine aircraft, should not lead to an accident - the design combined with the training of the crew should allow safe continuation of the flight. The same criteria apply to flight in adverse weather and when affected by human factors in either operations or maintenance.

High standards of flight crew training, proficiency and CRM can also serve to minimise the incorrect management by flight-crew of the onset of any in-flight reduction in airworthiness. A full understanding of the human factors issues involved in engineering and maintenance is therefore valuable.

The effective management of continuing airworthiness is an excellent defence. Defined as "all of the processes ensuring that, at any time in its operating life, the aircraft complies with the airworthiness requirements in force and is in a condition for safe operation". As part of continuing airworthiness management, each aircraft must hold a Certificate of Airworthiness (C of A) to prove that it conforms to the certificated Type Design and is in a condition for safe operation. In the EU, all aircraft must also be subject to a regular audit which leads to an airworthiness review certificate (ARC).

The regulator will require that the operator has in place a system to ensure compliance with the activities below. Some airworthiness authorities cover the following items with the term Certificate of Maintenance Review (CMR).

- Compliance with the maintenance programme.
- Embodiment of Mandatory Modifications and Inspections.
- Rectification of reported defects and investigation of adverse reliability matters.

How to Determine Airworthiness

Step 1: Determining Airworthiness

- pilot in command of a civil aircraft is responsible for determining whether that aircraft is in condition for safe flight.
    1. The pilot in command shall discontinue the flight when un-airworthy mechanical, electrical, or structural conditions occur.
- No person may operate any aircraft that has undergone maintenance, preventive maintenance, rebuilding, or alteration unless:
    1. It has been approved for return to service by a person authorized under Sec. 43.7 of this chapter; and
    2. The maintenance record entry required by Sec. 43.9 or Sec. 43.11, as applicable, has been made.
- No person may carry any person (other than crewmembers) in an aircraft that has been maintained, rebuilt, or altered in a manner that may have appreciably changed its flight characteristics or substantially affected its operation in flight until an appropriately rated pilot with at least a private pilot certificate flies the aircraft, makes an operational check of the maintenance performed or alteration made, and logs the flight in the aircraft records.
    1. The requirement for an in-flight operational check may be waived if, prior to flight, ground tests, inspection, or both show conclusively that the maintenance, preventive maintenance, rebuilding, or alteration has not appreciably changed the flight characteristics or substantially affected the flight operation of the aircraft.
- Aircraft must conform to the Type Certificate Data Sheet.

Step 2: Documents Required

- Aircraft must have documentation to show compliance with federal regulations.
- Those documents can be easily remembered through the acronym "ARROW:"
    1. A – Airworthiness Certificate
    2. R – Registration Certificate

3. R – Radio Certificate (international flights)
4. O – Operators Manual (specific for that aircraft) and placards (Compass deviation card/External data plate
5. W – Weight and Balance (current and specific to your airplane)

Step 3: Inspections Required

- In addition to the required documentation, inspections must be completed in accordance with the operation.
- The required inspections can be remembered with the acronym "AVIATE:"
    1. A – Airworthiness Directives
    2. V – VOR Check:
        - Required every 30 days for aircraft operating under Instrument Flight Rules
        - Governed by FAR 91.171
        - May be performed by any pilot
- I – Inspections:
    1. 100 hour:
        - Governed by FAR 91.409(b)
        - Aircraft operated for hire or used for flight instruction must have either an annual inspection or a 100-hour inspection every 100 hours of operation.
        - The 100-hour limit may be exceeded by no more that 10 hours for the sole purpose of flying to a location where the inspection can be completed.
        - Any excess time over the 100-hour due time will be subtracted from the next 100 hour inspection.
- Annual:
    1. Governed by FAR 91.409(a)
    2. Required for all aircraft
    3. Completed every 12 calendar months, expiring at the end of the month one year after the inspection
    4. Must be completed and properly endorsed by a mechanic with an inspection authorization (IA)
- A – Altimeter/Pitot-Static System (IFR Only)
    1. Altimeter inspections and tests are governed by Federal Aviation Regulation 91.411.
    2. Each altimeter must be tested and inspected within the preceding 24 calendar months

or following installation or maintenance on the automatic pressure altitude reporting system of the ATC transponder where data correspondence error could be introduced.

3. The static pressure system must be tested and inspected within the preceding 24 calendar months.

- These tests must be conducted by:
  1. The manufacturer of the airplane, or helicopter, on which the tests and inspections are to be performed;
  2. A certificated repair station
  3. A certificated mechanic with an airframe rating (static pressure system tests and inspections only)
  4. Altimeter and altitude reporting equipment approved under Technical Standard Orders are considered to be tested and inspected as of the date of their manufacture
  5. No person may operate an airplane, or helicopter, in controlled airspace under IFR at an altitude above the maximum altitude at which all altimeters and the automatic altitude reporting system of that airplane, or helicopter, have been tested.
- T – Transponder
  1. Transponder inspections and tests are governed by Federal Aviation Regulation 91.413
  2. Must be tested and inspected within the previous 24 months or after installation or maintenance where data correspondence error could be introduced
  3. Tests conducted by:
     - A certificated repair station
     - A holder of a continuous airworthiness maintenance program
     - The manufacturer of the aircraft on which the transponder to be tested is installed, if the transponder was installed by that manufacturer.
- E – Emergency Locator Transmitter (ELT)
  1. Must be tested and inspected every 12 calendar months
  2. Governed by FAR 91.207
  3. Must be replaced or recharged if the transmitter has been used for more than one hour of cumulative use
  4. Must be replaced or recharged if 50% of its useful life has expired.

Step 4: Required Equipment

- Visual Flight Rules:
  1. Day VFR (TOMATO FLAAMES):

- T - Tachometer (one for each engine)
- O - Oil Pressure Gauge (one for each liquid cooled engine)
- M - Magnetic Direction Indicator
- A - Airspeed indicator
- T - Temperature gauge (each liquid cooled engine)
- O - Oil temperature gauge (each liquid cooled engine)
- F - Fuel Gauges
- L - Landing gear position indicator lights (retractable landing gear)
- A - Altimeter
- A - Anti-Collision light system (strobes), if manufactured after 3/11/1996
- M - Manifold pressure gauge (each altitude engine)
- E - Emergency Locator Transmitter
- S - Seat belts & shoulder harness (front seats)

- Night VFR requires Day VFR plus (FLAPS)

1. F - Fuses (spares) or circuit breakers
2. L - Landing lights (if for hire)
3. A - Anti-collision lights
4. P - Position Lights
5. S - Source of energy (adequate to power all electrical and radio equipment)

- Instrument Flight Rules

1. Day/Night IFR (GRABCARD):
    - G - Generator or alternator
    - R - Rate of turn indicator (turn coordinator or turn & bank indicator)
    - A - Altimeter, sensitive, adjustable for barometric pressure (Kollsman window)
    - B - Ball (slip-skid indicator [inclinometer])
    - C - Clock (digital or analog displaying hours, minutes, and seconds)
    - A - Attitude indicator
    - R - Radios (radio communication and navigation equipment suitable for the route to be flown)
    - D - Direction (gyroscopic) indicator (directional gyro or heading indicator)

- Inoperative Equipment
    1. If any other item is inoperative, 14 CFR section 91.405 states that it shall be placarded as required by Sec. 43.11
    2. Following scheduled inspections:
        - Aircraft shall have discrepancies repaired unless it is permitted to be in operative by 91.213.
        - Maintenance personnel shall make appropriate entries in the aircraft maintenance records indicating the aircraft has been approved for return to service.
- Equipment impacting safety of flight must be repaired, but items that are not required may remain inoperative indefinitely provided they are appropriately placarded.

Step 5. Airworthiness Responsibilities

- Pilot In Command
    1. Determine airworthiness
- Owner:
    1. Register the aircraft
    2. Maintain the aircraft in an airworthy condition in accordance with FAR 47.403.

Step 6. Airworthiness Certificate

- On the actual airworthiness certificate itself, you will find:
    1. "Unless sooner surrendered, suspended, revoked, or a termination date is otherwise established by the Administrator, this airworthiness certificate is effective as long as the maintenance, preventive maintenance, and alterations are performed in accordance with parts 21, 43, and 91 of the federal aviation regulations, as appropriate, and the aircraft is registered in the United States".
- "Unless sooner surrendered, suspended, revoked, or a termination date is otherwise established by the Administrator".
    1. Various reasons including aircraft condition or legislation may null and void an airworthiness certificate.
- "This airworthiness certificate is effective as long as the maintenance, preventive maintenance, and alterations are performed in accordance with parts 21, 43, and 91 of the federal aviation regulations, as appropriate, and the aircraft is registered in the United States.
    1. The aircraft must be maintained in accordance with inspections listed above and preventative maintenance rules and regulations.

Chapter 2

Airframe Mechanic

The airframe of an aircraft refers to its mechanical structure consisting of the wings, fuselage and undercarriage. Airframe technicians are tasked with the inspection of aircraft frames, mechanical and electrical systems for detecting defects, wear and other issues in the airframe. They are also responsible for the repair and replacement of parts. This chapter discusses the vitals of repairing minor aircraft skin damage and fueling a plane, as well as the ways to become a successful airframe mechanic.

Airframe

Airframe is the basic structure of an airplane or spacecraft excluding its power plant and instrumentation; its principal components thus include the wings, fuselage, tail assembly, and landing gear. The airframe is designed to withstand all aerodynamic forces as well as the stresses imposed by the weight of the fuel, crew, and payload.

Most airframes of early airplanes consisted of a fuselage of truss design constructed of narrow hardwood boards or steel tubing and braced with wires. This basic framework supported the wing structure, which was composed of spars with ribbing. Both the fuselage and wings were covered by a skin of cotton fabric. Airframe construction was radically improved during the 1930s. The aerodynamically contoured fuselage shell characteristic of all modern aircraft was introduced at this time, and high-strength, lightweight metals (chiefly aluminum alloys, magnesium, and some stainless steel and titanium) replaced wood and fabric throughout the airframe.

The airframe of a fixed-wing aircraft consists of the following five major units:

1. Fuselage
2. Wings
3. Stabilizers
4. Flight controls surfaces
5. Landing gear

A rotary-wing aircraft consists of the following four major units:

1. Fuselage
2. Landing gear
3. Main rotor assembly
4. Tail rotor assembly

Fixed-Wing Aircraft

Fuselage

The fuselage is the main structure, or body, of the aircraft. It provides space for personnel, cargo, controls, and most of the accessories. The power plant, wings, stabilizers, and landing gear are attached to it.

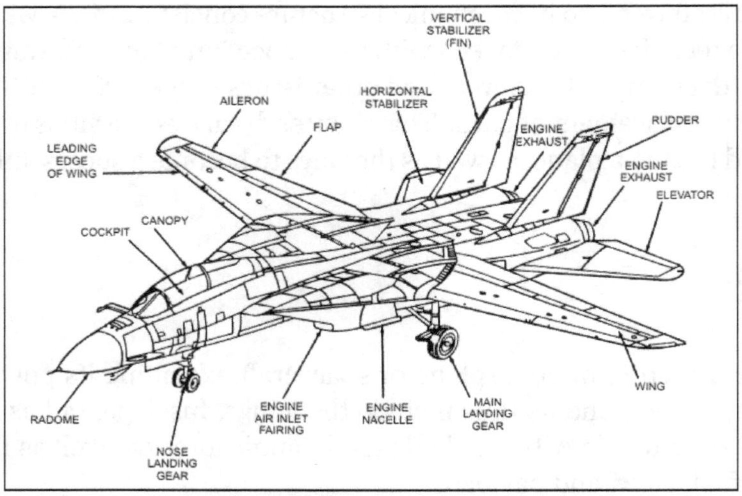

Principal Structural Units on an F-14 aircraft

There are two general types of fuselage construction—welded steel truss and monocoque designs. The welded steel truss was used in smaller Navy aircraft, and it is still being used in some helicopters.

The monocoque design relies largely on the strength of the skin, or covering, to carry various loads. The monocoque design may be divided into three classes—monocoque, semimonocoque, and reinforced shell.

- The true monocoque construction uses formers, frame assemblies, and bulkheads to give shape to the fuselage. However, the skin carries the primary stresses. Since no bracing members are present, the skin must be strong enough to keep the fuselage rigid. The biggest problem in monocoque construction is maintaining enough strength while keeping the weight within limits.
- Semimonocoque design overcomes the strength-to-weight problem of monocoque construction. In addition to having formers, frame assemblies, and bulkheads, the semimonocoque construction has the skin reinforced by longitudinal members.
- The reinforced shell has the skin reinforced by a complete framework of structural members. Different portions of the same fuselage may belong to any one of the three classes. Most are considered to be of semimonocoque-type construction.

The semimonocoque fuselage is constructed primarily of aluminum alloy, although steel and titanium are found in high-temperature areas. Primary bending loads are taken by the longerons, which usually extend across several points of support. The longerons are supplemented by other longitudinal members known as stringers. Stringers are more numerous and lightweight than longerons.

The vertical structural members are referred to as bulkheads, frames, and formers. The heavier vertical members are located at intervals to allow for concentrated loads. These members are also found at points where fittings are used to attach other units, such as the wings and stabilizers.

The stringers are smaller and lighter than longerons and serve as fill-ins. They have some rigidity but are chiefly used for giving shape and for attachment of skin. The strong, heavy longerons hold the bulkheads and formers. The bulkheads and formers hold the stringers. All of these join together to form a rigid fuselage framework. Stringers and longerons prevent tension and compression stresses from bending the fuselage.

The skin is attached to the longerons, bulkheads, and other structural members and carries part of the load. The fuselage skin thickness varies with the load carried and the stresses sustained at particular location.

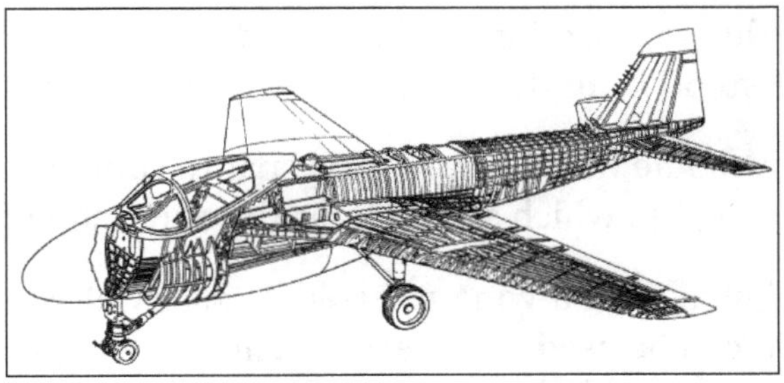

Semimonocoque fuselage construction.

There are a number of advantages in using the semimonocoque fuselage.

- The bulkhead, frames, stringers, and longerons aid in the design and construction of a streamlined fuselage. They add to the strength and rigidity of the structure.
- The main advantage of the semimonocoque construction is that it depends on many structural members for strength and rigidity. Because of its stressed skin construction, a semimonocoque fuselage can withstand damage and still be strong enough to hold together.

Points on the fuselage are located by station numbers. Station 0 is usually located at or near the nose of the aircraft. The other stations are located at measured distances (in inches) aft of station 0. A typical station diagram is shown in figure. On this particular aircraft, fuselage station (FS) 0 is located 93.0 inches forward of the nose.

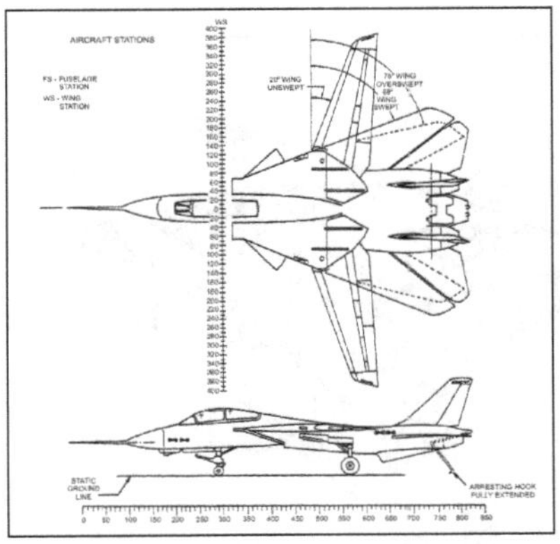

Fuselage station diagram of an F-14 aircraft.

Wings

Wings develop the major portion of the lift of a heavier-than-air aircraft. Wing structures carry some of the heavier loads found in the aircraft structure. The particular design of a wing depends on many factors, such as the size, weight, speed, rate of climb, and use of the aircraft. The wing must be constructed so that it holds its aerodynamics shape under the extreme stresses of combat maneuvers or wing loading.

Wing construction is similar in most modern aircraft. In its simplest form, the wing is a framework made up of spars and ribs and covered with metal. The construction of an aircraft wing is shown in figure.

Spars are the main structural members of the wing. They extend from the fuselage to the tip of the wing. All the load carried by the wing is taken up by the spars. The spars are designed to have great bending strength. Ribs give the wing section its shape, and they transmit the air load from the wing covering to the spars. Ribs extend from the leading edge to the trailing edge of the wing

In addition to the main spars, some wings have a false spar to support the ailerons and flaps. Most aircraft wings have a removable tip, which streamlines the outer end of the wing.

Most Navy aircraft are designed with a wing referred to as a wet wing. This term describes the wing that is constructed so it can be used as a fuel cell. The wet wing is sealed with a fuel-resistant compound as it is built. The wing holds fuel without the usual rubber cells or tanks.

The wings of most naval aircraft are of all metal, full cantilever construction. Often, they may be folded for carrier use. A full cantilever wing structure is very strong. The wing can be fastened to the fuselage without the use of external bracing, such as wires or struts.

A complete wing assembly consists of the surface providing lift for the support of the aircraft. It also provides the necessary flight control surfaces.

Stabilizers

The stabilizing surfaces of an aircraft consist of vertical and horizontal airfoils. They are called the vertical stabilizer (or fin) and horizontal stabilizer. These two airfoils, along with the rudder and elevators, form the tail section. For inspection and maintenance purposes, the entire tail section is considered a single unit called the empennage.

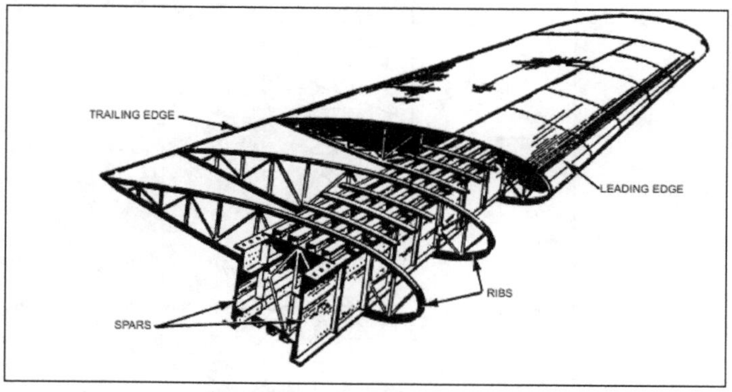

Two-spar wing construction.

The main purpose of stabilizers is to keep the aircraft in straight-and-level flight. The vertical stabilizer maintains the stability of he aircraft about its vertical axis. This is known as directional stability. The vertical stabilizer usually serves as the base to which the rudder is attached. The horizontal stabilizer provides stability of the aircraft about its lateral axis. This is known as longitudinal stability. The horizontal stabilizer usually serves as the base to which the elevators are attached. On many newer, high-performance aircraft, the entire vertical and/or horizontal stabilizer is a movable airfoil. Without the movable airfoil, the flight control surfaces would lose their effectiveness at extremely high altitudes.

Stabilizer construction is similar to wing construction. For greater strength, especially in the thinner airfoil sections typical of trailing edges, a honeycomb-type construction is used. Some larger carrier-type aircraft have vertical stabilizers that are folded hydraulically to aid aircraft movement aboard aircraft carriers.

Flight Control Surfaces

Flight control surfaces are hinged (movable) airfoils designed to change the attitude of the aircraft during flight. These surfaces are divided into three groups—primary, secondary, and auxiliary.

Primary Group

The primary group of flight control surfaces includes ailerons, elevators, and rudders. The ailerons attach to the trailing edge of the wings. They control the rolling (or banking) motion of the aircraft. This action is known as longitudinal control.

The elevators are attached to the horizontal stabilizer and control the climb or descent (pitching motion) of the aircraft. This action is known as lateral control.

The rudder is attached to the vertical stabilizer. It determines the horizontal flight (turning or yawing motion) of the aircraft. This action is known as directional control.

The ailerons and elevators are operated from the cockpit by a control stick on single-engine aircraft. A yoke and wheel assembly operates the ailerons and elevators on multiengine aircraft, such as transport and patrol aircraft. The rudder is operated by foot pedals on all types of aircraft.

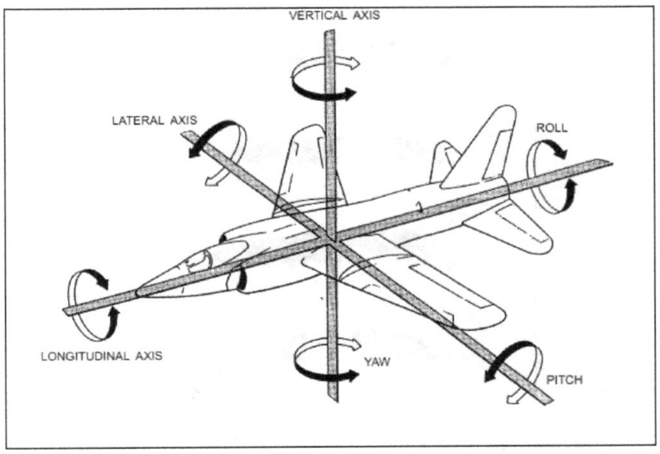

Axes and fundamental movements of the aircraft.

Secondary Group

The secondary group includes the trim tabs and spring tabs. Trim tabs are small airfoils recessed into the trailing edges of the primary control surface. Each trim tab hinges to its parent primary control surface, but operates by an independent control. Trim tabs let the pilot trim out an unbalanced condition without exerting pressure on the primary controls.

Spring tabs are similar in appearance to trim tabs but serve an entirely different purpose. Spring tabs are used for the same purpose as hydraulic actuators. They aid the pilot in moving a larger control surface, such as the ailerons and elevators.

Auxiliary Group

The auxiliary group includes the wing flaps, spoilers, speed brakes, and slats.

WING FLAPS.—Wing flaps give the aircraft extra lift. Their purpose is to reduce the landing speed. Reducing the landing speed shortens the length of the landing rollout. Flaps help the pilot land in small or obstructed areas by increasing the glide angle without greatly increasing the approach speed. The use of flaps during takeoff serves to reduce the length of the takeoff run.

Some flaps hinge to the lower trailing edges of the wings inboard of the ailerons. Leading edge flaps are used on the F-14 Tomcat and F/A-18 Hornet. Four types of flaps are shown in figure. The plain flap forms the trailing edge of the airfoil when the flap is in the up position. In the split flap, the trailing edge of the airfoil is split, and the lower half is hinged and lowers to form the flap. The fowler flap operates on rollers and tracks, causing the lower surface of the wing to roll out and then extend downward. The leading edge flap operates like the plain flap. It is hinged on the bottom side. When actuated, the leading edge of the wing actually extends in a downward direction to increase the camber of the wing. Landing flaps are used in conjunction with other types of flaps.

SPOILERS.—Spoilers are used to decrease wing lift. The specific design, function, and use vary with different aircraft. On some aircraft, the spoilers are long narrow surfaces, hinged at their leading edge to the upper surfaces of the wings. In the retracted position, they are flush with the wing skin. In the raised position, they greatly reduce wing lift by destroying the smooth flow of air over the wing surface.

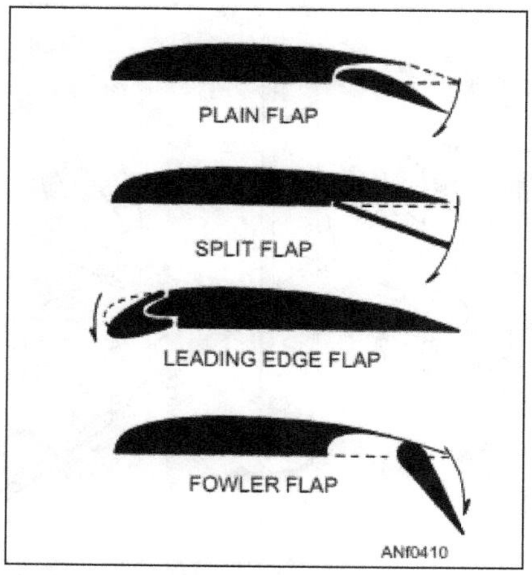

Types of flaps.

SPEED BRAKES.—Speed brakes are movable control surfaces used for reducing the speed of the aircraft. Some manufacturers refer to them as dive brakes; others refer to them as dive flaps. On some aircraft, they're hinged to the sides or bottom of the fuselage. Regardless of their location, speed brakes serve the same purpose—to keep the airspeed from building too high when the aircraft dives. Speed brakes slow the aircraft's speed before it lands.

SLATS.—Slats are movable control surfaces that attach to the leading edge of the wing. When the slat is retracted, it forms the leading edge of the wing. When the slat is open (extended forward), a slot is created between the slat and the wing leading edge. High-energy air is introduced into the boundary layer over the top of the wing. At low airspeeds, this action improves the lateral control handling characteristics. This allows the aircraft to be controlled at airspeeds below normal landing speed. The high-energy air that flows over the top of the wing is known as boundary layer control air. Boundary layer control is intended primarily for use during operations from carriers. Boundary layer control air aids in catapult takeoffs and arrested landings. Boundary control air can also be accomplished by directing high-pressure engine bleed air across the top of the wing or flap surface.

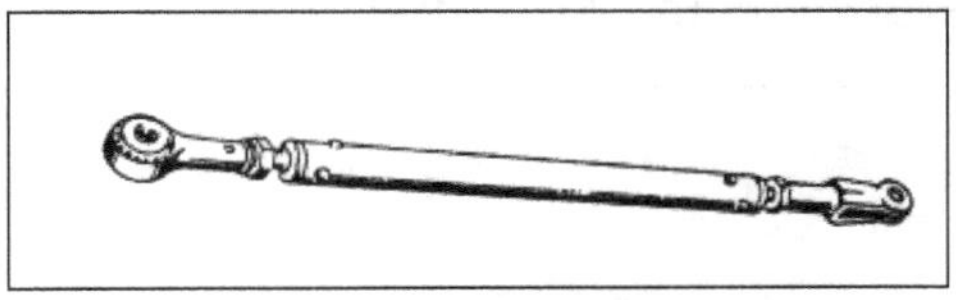

Push-pull tube assembly

Flight Control Mechanisms

The term flight control refers to the linkage that connects the control(s) in the cockpit with the flight control surfaces. There are several types of flight controls in naval aircraft; some are manually operated while others are power operated.

Manually operated flight control mechanisms are further divided into three groups—cable operated, push-pull tube operated, and torque tube operated. Some systems may combine two or more of these types.

In the manually operated cable system, cables are connected from the control in the cockpit to a bell crank or sector. The bell crank is connected to the control surface. Movement of the cockpit controls transfers force through the cable to the bell crank, which moves the control surface.

In a push-pull tube system, metal push-pull tubes (or rods) are used as a substitute for the cables. Push-pull tubes get their name from the way they transmit force.

In the torque tube system, metal tubes (rods) with gears at the ends of the tubes are used. Motion is transmitted by rotating the tubes and gears.

On all high-performance aircraft, the control surfaces have great pressure exerted on them. At high airspeed, it is physically impossible for the pilot to move the controls manually. As a result, power-operated control mechanisms are used. In a power-operated system, a hydraulic actuator (cylinder) is located within the linkage to assist the pilot in moving the control surface.

A typical flight control mechanism is shown in figure. This is the elevator control of a lightweight trainer-type aircraft. It consists of a combination of push-pull tubes and cables.

The control sticks in the system shown in figure are connected to the forward sector by push-pull tubes. The forward sector is connected to the aft (rear) sector by means of cable assemblies. The aft sector is connected to the flight control by another push-pull tube assembly.

Landing Gear

Before World War II, aircraft were made with their main landing gear located behind the center of gravity. An auxiliary gear under the fuselage nose was added. This arrangement became known as the tricycle type of landing gear. Nearly all present-day Navy aircraft are equipped with tricycle landing gear. The tricycle gear has the following advantages over older landing gear:

- More stable in motion on the ground
- Maintains the fuselage in a level position
- Increases the pilot's visibility and control
- Makes landing easier, especially in cross wind

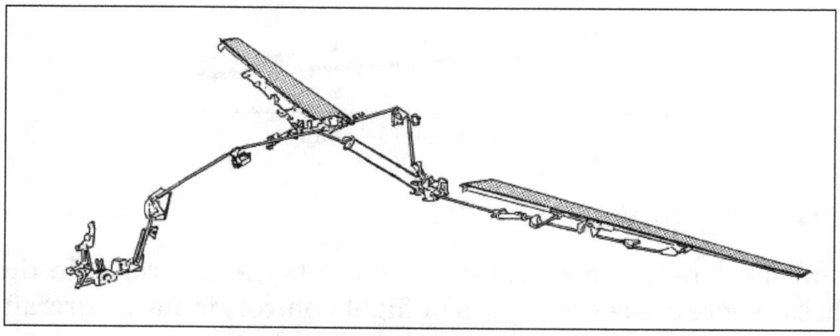

Typical flight control mechanism.

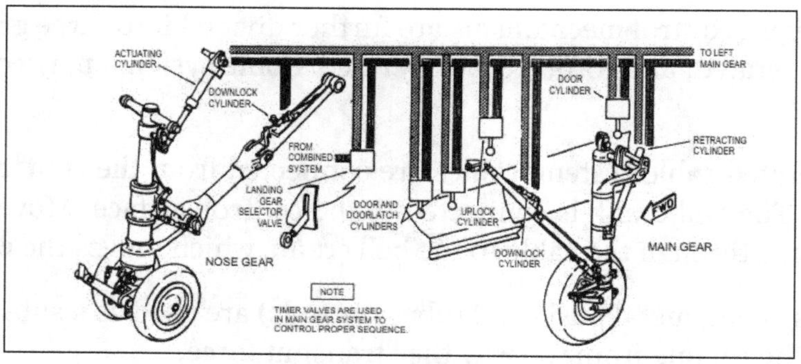

Typical landing gear system.

The landing gear system consists of three retractable landing gear assemblies. Each main landing gear has a conventional air-oil shock strut, a wheel brake assembly, and a wheel and tire assembly. The nose landing gear has a conventional air-oil shock strut, a shimmy damper, and a wheel and tire assembly.

The shock strut is designed to absorb the shock that would otherwise be transmitted to the airframe during landing, taxiing, and takeoff. The air-oil strut is used on all naval aircraft. This type of strut has two telescoping cylinders filled with hydraulic fluid and compressed air or nitrogen. Figure shows the internal construction of one type of air-oil shock strut.

The main landing gear is equipped with brakes for stopping the aircraft and assisting the pilot in steering the aircraft on the ground.

The nose gear of most aircraft can be steered from the cockpit. This provides greater ease and safety on the runway when landing and taking off and on the taxiway in taxiing.

Arresting Gear

A carrier-type aircraft is equipped with an arresting hook for stopping the aircraft when it lands on the carrier. The arresting gear has an extendible hook and the mechanical, hydraulic, and pneumatic equipment necessary for hook operation. The arresting hook on most aircraft releases mechanically, lowers pneumatically, and raises hydraulically.

The hook hinges from the structure under the rear of the aircraft. A snubber meters hydraulic fluid and works in conjunction with nitrogen pressure. The snubber holds the hook down and prevents it from bouncing when it strikes the carrier deck.

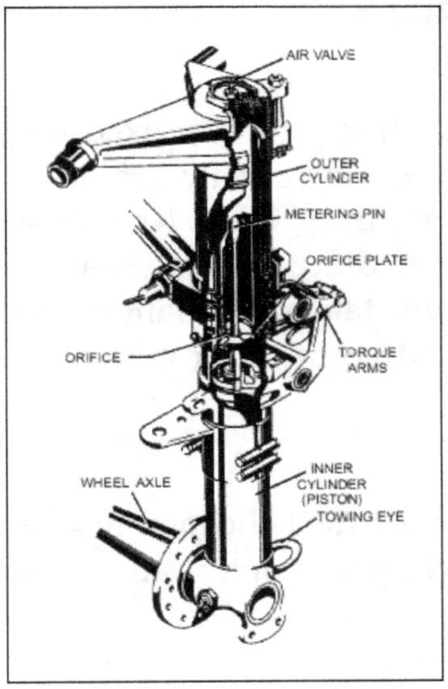

Internal construction of a shock strut.

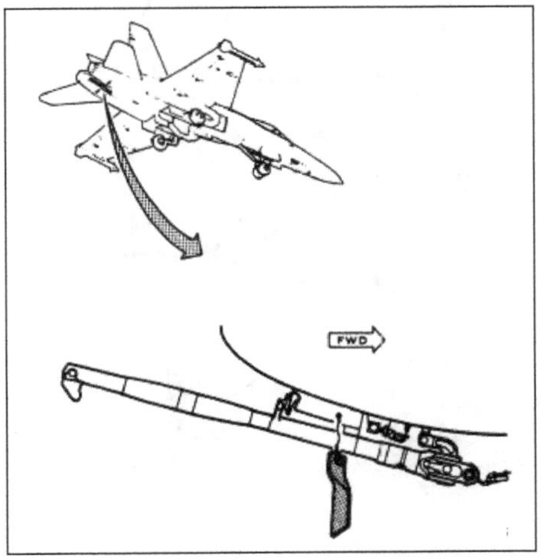

Arresting gear installation.

Catapult Equipment

Carrier aircraft have built-in equipment for catapulting off the aircraft carrier. Older aircraft had hooks on the airframe that attached to the cable bridle. The bridle hooks the aircraft to the ship's catapult. Newer aircraft have a launch bar built into the nose landing gear assembly. The holdback assembly allows the aircraft to be secured to the carrier deck for full-power turnup of the engine prior to takeoff. For nose gear equipment, a track attaches to the deck to guide the nosewheel into position. The track has provisions for attaching the nose gear to the catapult shuttle and for holdback.

Rotary-Wing Aircraft

Within the past 20 years, helicopters have become a reality, and are found throughout the world. They perform countless tasks suited to their unique capabilities.

A helicopter has one or more power-driven horizontal airscrews (rotors) to develop lift and propulsion. If a single main rotor is used, it is necessary to employ a means to counteract torque. If more than one main rotor (or tandem) is used, torque is eliminated by turning each main rotor in opposite directions.

The fundamental advantage the helicopter has over fixed-wing aircraft is that lift and control are independent of forward speed. A helicopter can fly forward, backward, or sideways, or it can remain in stationary flight (hover) above the ground. No runway is required for a helicopter to take off or land. For example, the roof of an office building is an adequate landing area. The helicopter is considered a safe aircraft because the takeoff and landing speed is zero, and it has autorotational capabilities. This allows a controlled descent with rotors turning in case of engine failure in flight.

Fuselage

Like the fuselage of a fixed-wing aircraft, the helicopter fuselage may be welded truss or some form of monocoque construction. Many Navy helicopters are of the monocoque design.

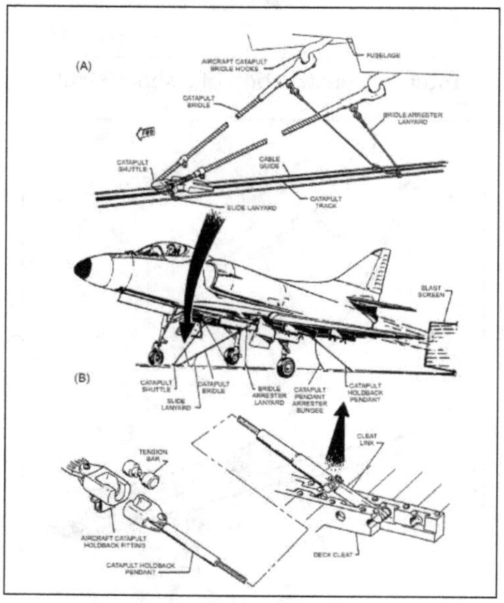

Aircraft catapult equipment.

A typical Navy helicopter, the H-60, is shown in figure. Some of its features include a single main rotor, twin engine, tractor-type canted tail rotor, controllable stabilizer, fixed landing gear, rescue hoist, external cargo hook, and weapons pylons. The fuselage consists of the entire airframe, sometimes known as the body group.

The body group is an all-metal semimonocoque construction. It consists of an aluminum and titanium skin over a reinforced aluminum frame.

Landing Gear Group

The landing gear group includes all the equipment necessary to support the helicopter when it is not in flight. There are several types of landing gear on helicopters—conventional fixed (skid type), retractable, and nonretractable.

Main Landing Gear

The H-60's nonretracting main landing gear consists of two single axle, air/oil type of shock-strut assemblies that mount to the fuselage. Each is equipped with tubeless tires, hydraulic disc brakes, tie-down rings, drag braces, and safety switches. They are part of the lower end of the shock strut piston.

Tail Landing Gear

The H-60's tail landing gear is a nonretracting, dual wheel, 360-degree swiveling type. It is equipped with tubeless tires, tie-down ring, shimmy damper, tail-wheel lock, and an air/oil shock-strut, which serves as an aft touchdown point for the pilots to cushion the landing shock.

Main Rotor Assembly

The main rotor (rotor wing) and rotor head (hub assembly) are identical in theory of flight but differ in engineering or design. They are covered here because their functions are closely related. The power plant, transmission, drive-train, hydraulic flight control, and rotor systems all work together. Neither has a function without the other.

Rotary Wing

The main rotor on the H-60 has four identical wing blades. Other types of helicopters may have two, four, five, six, or seven blades. Figure shows some typical rotor blades.

H-60 helicopter.

Rotary-wing blades are made of titanium, aluminum alloys, fiber glass, graphite, honeycomb core, nickel, and steel. Each has a nitrogen-filled, pressurized, hollow internal spar, which runs the length of the blade. The cuff provides the attachment of the blade to the rotor hub. A titanium abrasion strip covers the entire leading edge of the spar from the cuff end to the removable blade tip faring. This extends the life of the rotor blade.

The examples shown in figure show other features—trim tabs, deicing protection, balance markings, and construction.

Main Rotor Head/Hub Assembly

The rotor head is fully articulating and is rotated by torque from the engines through the drive train and main gearbox or transmission. The flight controls and hydraulic servos transmit movements to the rotor blades. The principal components of the rotor head are the hub and swashplate assemblies. The hub is one piece, made of titanium and sits on top of the rotor mast. Attaching components are the sleeve and spindles, blade fold components, vibration absorber, bearings, blade dampers, pitch change horns, adjustable pitch control rods, blade fold hinges, balance weights, antiflapping and droop stops, and faring.

The swashplate consists of a rotating disc (upper), stationary (lower) portion with a scissors and sleeve assembly separated by a bearing. The swashplate is permitted to slide on the main rotor vertical driveshaft and mounts on top the main transmission. The entire assembly can tilt in any direction following the motion of the flight controls.

The hydraulic servo cylinders, swashplate, and adjustable pitch control rods permit movement of the flight controls to be transmitted to the rotary-wing blades. The sleeve and spindle and blade dampers allow limited movement of the blades in relation to the hub. These movements are known as lead, lag, and flap.

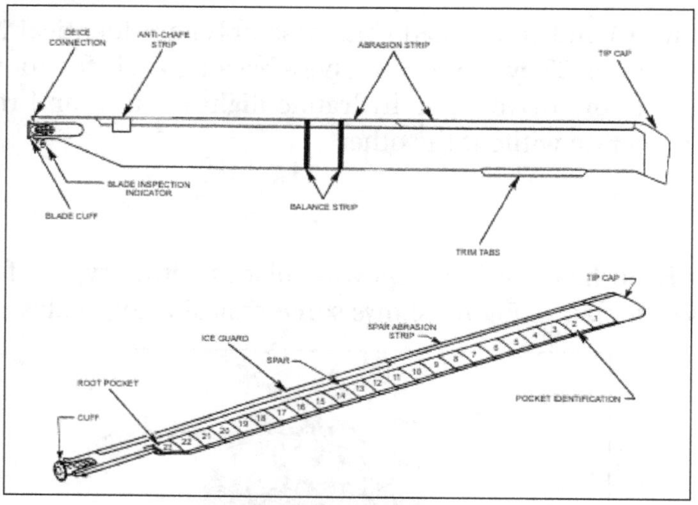

Types of main rotor blades.

- Lead occurs during slowing of the drive mechanism when the blades have a tendency to
- remain in motion.

- Lag is the opposite of lead and occurs during acceleration when the blade has been at rest and tends to remain at rest.
- Flap is the tendency of the blade to rise with high-lift demands as it tries to screw itself upward into the air.

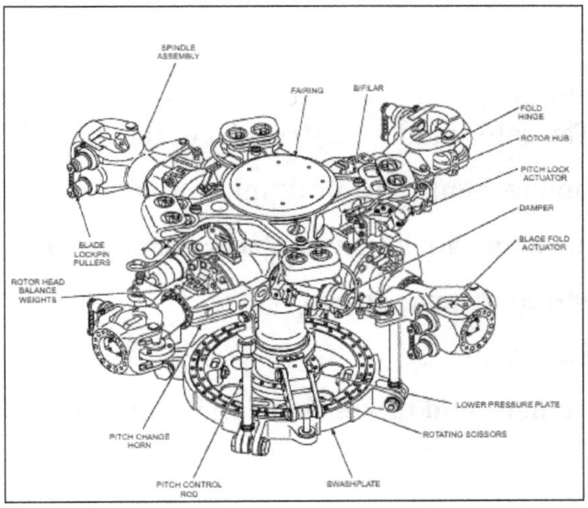

Main rotor head/hub assembly.

Antiflapping stops and droop stops restrict flapping and conning motion of the rotary-wing head and blades at low rotor rpm when slowing or stopping.

Tail Rotor Group

The directional control and antitorque action of the helicopter is provided by the tail rotor group. See blade fold actuator fairing bifilar damper pitch lock actuator rotor hub fold hinge spindle assembly blade lockpin pullers rotating scissors lower pressure plate pitch control swashplate rod pitch change horn anf0119 rotor head balance weights figure. Main rotor head/hub assembly. These components are similar in function to the main rotor.

Pylon

The pylon, shown in figure, attaches on the aircraft to the main fuselage by hinge fittings. These hinge fittings serve as the pivot point for the pylon to fold along the fuselage. Folding the pylon reduces the overall length of the helicopter, which helps for confined shipboard handling.

The pylon houses the intermediate and tail rotor gearboxes, tail rotor drive shaft, cover, tail bumper, position/anticollision lights, hydraulic servos, flight control push-pull tubes/cables/bell cranks, stabilizer/ elevator flight control surface, some antennas, and rotary rudder assembly.

Rotary Rudder Head

The rudder head can be located on either side of the pylon, depending on the type of aircraft, and includes such items as the hub, spindle, pitch control beam, pitch change links, bearings, and tail rotor blades.

Change in blade pitch is accomplished through the pitch change shaft that moves through the horizontal shaft of the tail gearbox, which drives the rotary rudder assembly. As the shaft moves inward toward the tail gearbox, pitch of the blade is decreased. As the shaft moves outward from the tail gearbox, pitch of the blade is increased. The pitch control beam is connected by links to the forked brackets on the blade sleeves.

Rotary Rudder Blades

Like the blades on a main rotor head, the blades found on a rotary rudder head may differ, depending on the type of aircraft. Tail rotor blades may consist of the following components:

- Aluminum alloy, graphite composite, or titanium spar
- Aluminum pocket and skin with honeycomb core or cross-ply fiber glass exterior
- Aluminum or graphite composite tip cap
- Aluminum trailing edge cap
- Aluminum or polyurethane and nickel abrasion leading edge strip

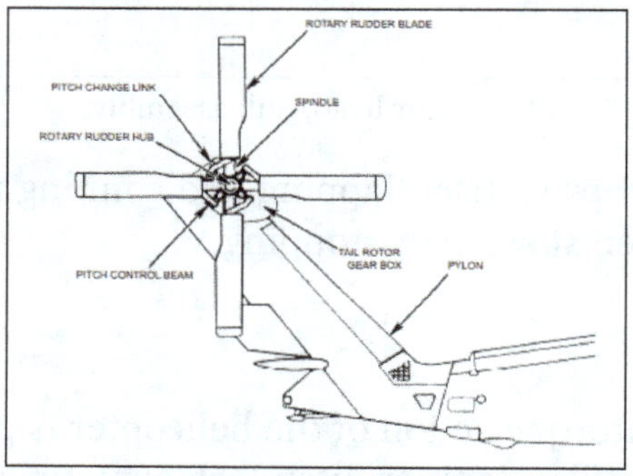

Tail rotor group.

Additionally, rotary rudder blades may have deicing provisions, such as electrothermal blankets that are bonded into the blade's leading edge. or a neoprene anti-icing guard embedded with electrical heating elements.

Airframe Mechanic

Airframe technicians are so-called because most employers prefer to hire those with airframe and power plant certifications. They are more commonly known as aircraft mechanics and are classified as such by the Standard Occupational Code of the Bureau of Labor Statistics. Their primary responsibility is to keep aircraft in working order.

Airframe technicians run thorough inspections of aircraft frames, mechanical components and electrical systems to locate wear, defects and other problems. They read documentation and other repair manuals to determine standards and procedures, and test aircraft functions using diagnostic instruments to ensure those standards are being met. They then repair or replace components using hand or power tools and may order additional supplies. Meticulous records of all activities are needed, which can then be presented to managers or inspectors when requested.

Types

Because of the different types of aircraft needing servicing, airframe technicians often specialize in such categories as passenger jetliners, propeller-driven airplanes or helicopters. They may also focus on different systems, such as engines or hydraulics. The BLS divides technicians into two main categories -- maintenance and repair mechanics. Maintenance mechanics specialize in preventative maintenance, periodically inspecting aircraft based on a combination of the hours flown and the days since the last inspection. Aside from visual examination, they can rely on automatic monitoring systems on aircraft that have them. Repair mechanics focus on repairing systems by fixing problems that inspections reveal or that pilots report. Both types typically work under strict deadlines, which may cause stress.

Safety

The hangars, repair stations and airfields in which technicians work can be dangerous for the inattentive because of the powerful, high-speed aircraft they are around. In addition, technicians may handle dangerous chemicals or lift heavy objects, sometimes from scaffolds and ladders. These factors contribute to a higher-than-average rate of injury and illness for the profession, according to the BLS. Technicians minimize hazards by observing strict safety procedures and wearing safety gear such as headphones to counter the noise generated by engines. Most work in eight-hour day, evening and night shifts, with the best schedules going to those with the most seniority.

How to Become an Airframe Mechanic

Airframe mechanics, like other types of aircraft mechanics, usually work full-time, with overtime and weekend work as a common requirement. Such mechanics typically work on aircraft in hangars, repair stations, or even on airfields. They work as a team, and this job can be physically demanding. Airframe mechanics must be comfortable working in small, uncomfortable spaces and with heights.

Career Requirements

Degree Level	Associate's or bachelor's degree
Degree Field(s)	Aviation maintenance or aviation technology
Certification	Must be certified or must work under the supervision of a certified mechanic
Key Skills	Physical agility; attention to detail; manual dexterity; familiarity with hand and power tools, welding tools, and gauges
Salary	$58,370 (2015 median for all aircraft mechanics and service technicians)

Complete a Training Program

There are three paths to fulfilling FAA requirements as an airframe mechanic. One is to graduate from an approved aviation maintenance technician program. These programs may take two to four years to complete, and a high school diploma or its equivalent is necessary for admission. A second option is to be trained under the supervision of a certified airframe mechanic for a minimum of 18 months. The last is to serve in the military and obtain FAA-approved airframe maintenance training as a military occupational specialty.

Many airframe mechanics are also trained in aircraft powerplants, which means learning to repair and maintain engines and turbines. Many employers prefer applicants who have Airframe & Powerplant (A&P) certification from the FAA.

Pass Airframe Certification Exams

The FAA requires all certified airframe mechanics to successfully complete written, oral, and practical examinations. Candidates must be 18 years of age with the ability to read, write, and speak English. Those who haven't completed an FAA-approved training program must have at least 18 months of hands-on airframe experience.

Applicants present proof of experience or training to a local FAA office and then are directed to an approved computer testing facility for the multiple-choice written exam. An FAA-designated mechanic examiner administers the oral and practical examinations. Applicants must pass all of the tests within a 24-month period in order to be eligible for certification.

Apply for a Position

Mechanics are often hired by aerospace parts manufacturers, commercial airlines, airports, and government agencies. In addition to examining, fixing, and testing aircraft parts, airframe mechanics must document all service and repairs. Most of these professionals work full time, though overtime may be required.

Airframe mechanic certification is valid for two years. To maintain this credential, mechanics must complete one refresher class in that two-year period and perform at least one inspection or repair every 90 days.

Airframe mechanics can advance in the field both through work experience and education. In order to work as an aircraft inspector, a mechanic must have held A&P certification for three years and must have at least two years of relevant work experience. Additionally, pursuing a bachelor's degree in a related subject, such as avionics or aviation maintenance, can lead to advancement in the field.

How to Repair Minor Aircraft Skin Damage

This procedure shows how to make a simple repair to aircraft skin for example to repair a dent caused by dropping a tool on the aluminum surface. The time required varies but typically take 1-3 hours depending on skill level and the severity of the damage. The reader should be familiar working with sheet metal and riveting.

Steps

1. Draw a circle around the damaged area.

2. Drill a pilot hole in the circle so a bigger Unibit can easily drill into the aluminum.

3. Using a preferred method cut out the aluminum within the circle drawn around the damage. Debur the newly cut hole.

4. Cut out a circular patch from a sheet of aluminum that is the same thickness and the original aircraft skin. The diameter will depend on the number of rows of rivets you're planning on installing. Debur the outer edge of the patch.

5. Center the patch over the hole in the aircraft skin and mark its outline.

6. Since the aluminum being used here is 0.040" thick, and the rivets are 4/32" in diameter. The edge distance for riveting will be 1/4". The patch cut for this repair is big enough in diameter for 2 rows of rivets, so you will draw a line 1/4" from the small hole and 1/4" inward from the outline of the patch you drew.

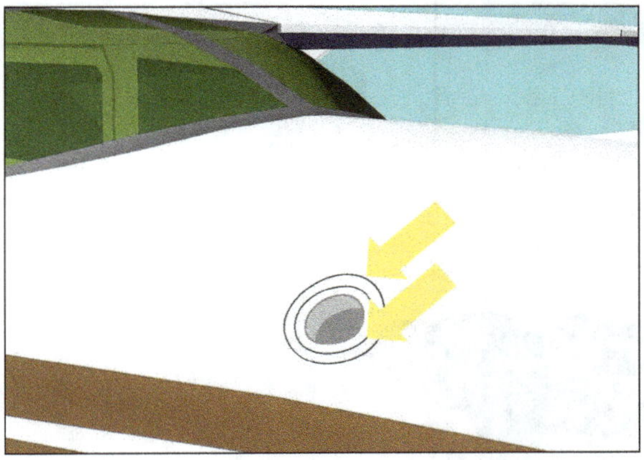

7. Once the edge distance lines are drawn, it's time to mark the location of each rivet. The rule for rivet spacing is a minimum of 4 times the diameter of the rivet and a maximum of 12 times the diameter of the rivet. You'll also want to keep the spacing between each rivet the same. Just make sure you're within those limits.

8. Next, drill pilot holes where the rivet makings are located with a small drill bit so it is possible to drill the holes to full size, without deformation of the aluminum sheet.

9. Secure the patch on the aircraft skin and drill pilot holes in the patch. Use the holes previously drilled in the skin as a guide.

10. With the patch still secured to the aircraft skin, expand the holes to the proper diameter. Use clecos to make sure the patch does not move while drilling.

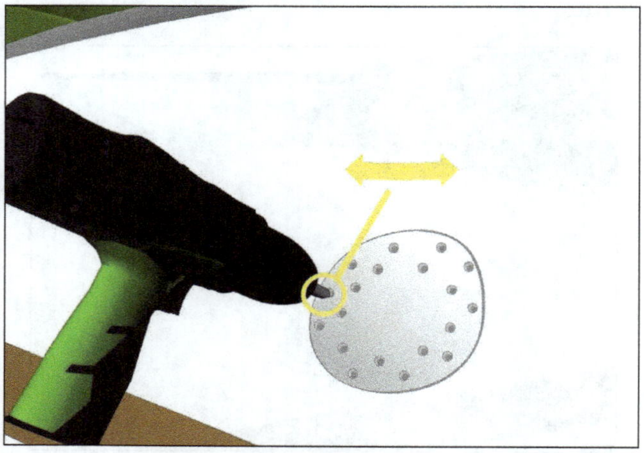

11. Remove the patch from the aircraft skin and deburr all the holes both on the patch and on the aircraft skin.

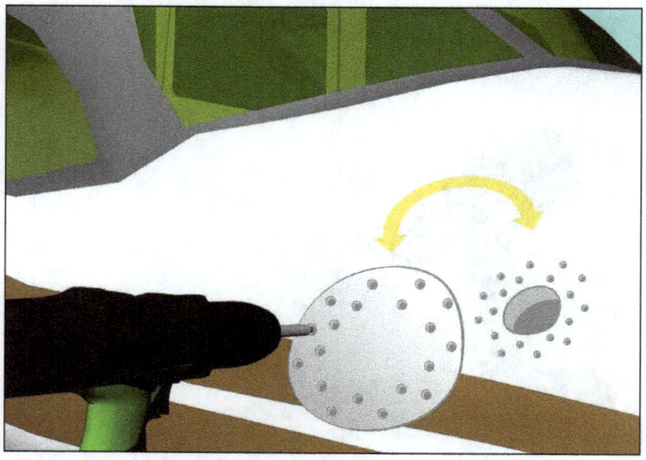

12. Once deburred, re-secure the patch on the skin using clecos.

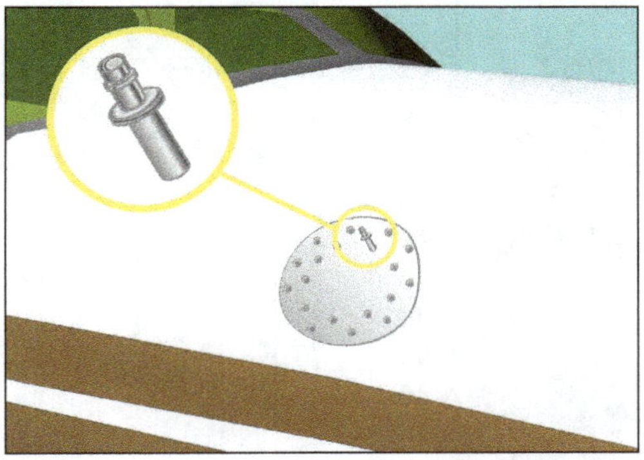

13. Using the proper rivets, secure the patch to the aircraft skin.

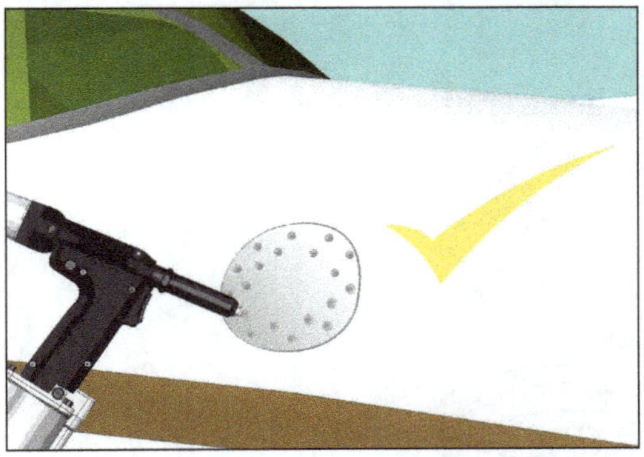

How to Fuel a Plane

Ever wanted to know how to fuel a plane? Fuelling a plane is not the same as filling up your car. By following the steps outlined in the following instructions, you will have the necessary knowledge to be able to refuel an airplane safely.

Steps

1. Determine what fuel grade is required for your aircraft.

2. Determine how much fuel is required.

3. Connect the ground wire to the aircraft

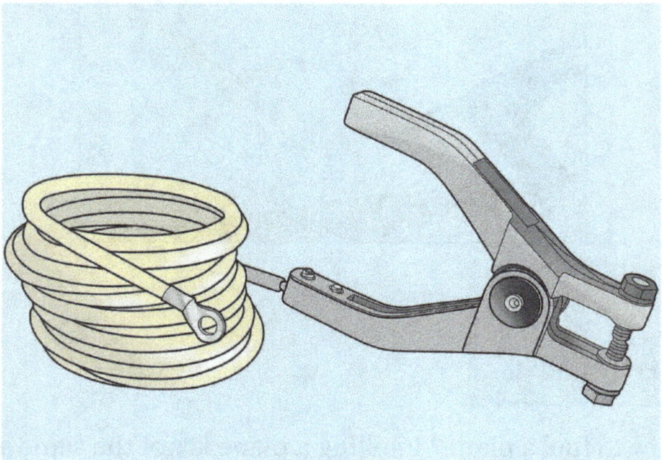

4. For high wing aircraft put ladder in front of the fuel cap on the furthest fuel tank.

5. Drag out fuel hose, for as far as you need it.

6. Remove fuel cap.

7. As you see in the video, this is not like fuelling your car. On larger aircraft, you must fill up the wings first, simply for the balance of the plane. While fuelling make sure all selected valves are correctly opened. If they are not check with the pilot or contact maintenance.

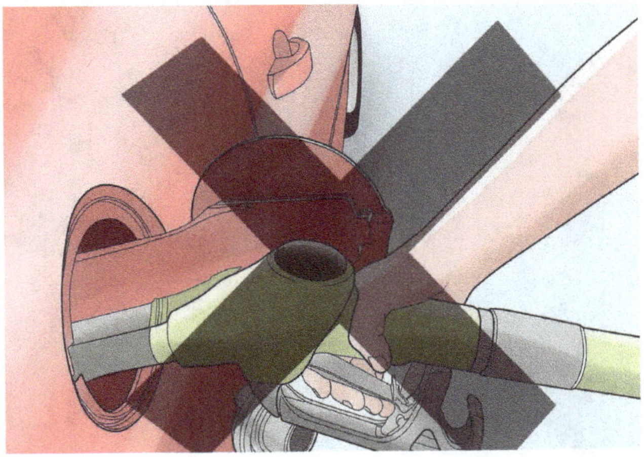

8. You must fill up the wing tanks before fuelling the centre tank.

9. After fuelling reinstall fuel cap and close the panel correctly.

10. Wind up fuel hose.

11. Wind up ground wire.

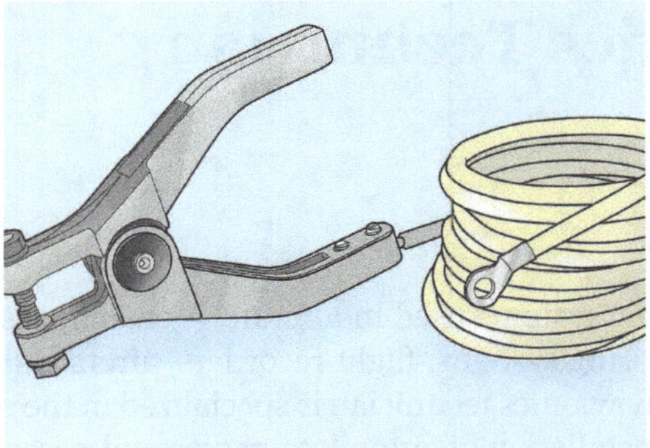

12. Double check fuel caps are tight.

Chapter 3 Avionics Technician

Avionics refer to the electronic systems used in aircraft. It encompasses navigation systems, communications systems, monitoring systems, flight recorders, aircraft flight-control systems, aircraft management systems, etc. An avionics technician is specialized in the maintenance of all these electronic components and the installation of autopilots, radios and passenger entertainment systems. The aim of this chapter is to explore the various aspects of radio communication, aircraft weight and balance, autopilot systems, antennas, etc. and the various responsibilities of avionics technicians.

Avionics

Essentially, avionics refer to aviation electronics. Demand exists for well-trained technicians, commonly called aircraft mechanics, who can maintain, troubleshoot, and repair aviation electronics.

Avionics, or aviation electronics, is a general term for all electrical equipment used in missiles, aircraft and spacecraft. Avionics encompasses components the pilot directly uses, such as navigation, weather radar, and radio communication equipment. In addition, avionics also comprises other electronic systems not directly used by the pilot that control and monitor flight and engine performance.

Avionics Technician

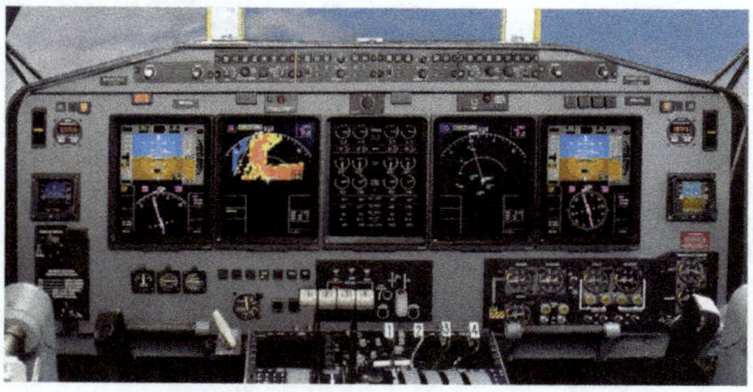

What does an Avionics Technician do

Some avionics technicians install and troubleshoot avionics equipment on fixed wing aircraft or helicopters. They're responsible to see that everything works properly and none of it interferes with other electronic devices on board. When a device fails, they remove it and send it to a shop where a bench avionics technician performs repairs. In a small facility, a technician may work on the aircraft and in the shop, but in large operations the jobs are separate.

Bench technicians may work on navigation and communication radios, autopilots, computers, and even large mechanical assemblies like radar antennas. A large shop may work on hundreds of different items. Avionics technicians are often specialists within one or two areas. The job requires excellent troubleshooting skills and the ability to do intricate, high-reliability soldering on delicate components. Good bench technicians have patience, steady hands, and excellent vision. Modern circuit boards are densely populated with tiny integrated circuits. Much of the soldering is done with binocular magnifiers or microscopes.

System trouble-shooters are avionics technicians with years of experience and a tremendous depth of knowledge about commercial aircraft. They may not know the intricacies of the individual components that make up the system, but they know how those parts interact and sometimes counteract one another. Line avionics technicians rely on them for accurate diagnosis of difficult problems. There's always a deadline. An aircraft has to fly again shortly, so it seems that these trouble-shooters are constantly on the run.

Workplace of an Avionics Technician

Avionics technicians are employed in several different settings. Most passengers inside a terminal have seen a line technician rushing to replace a black box so that a flight can leave on time, but the bulk of the work around a terminal is done at night. Pilots write up problems in the aircraft log book during the day. Avionics technicians try to clear those entries at night. They may be updating aircraft software, repairing wiring, or changing black boxes as necessary. The night crew technicians repair those time consuming items that could be deferred overnight, provided that safety is not compromised.

Hangar avionics technicians have the luxury of time, unlike the line technicians. Their airplane may be in the hangar for days or weeks. A planner reviews the aircraft records and any new information from the manufacturer, and then works out a schedule to cover all the necessary routine and non-routine maintenance. Teams are assigned and the time required for each task is included on the schedule. Unanticipated problems, like a crushed antenna cable, can introduce delays, but the teams usually manage to get the work finished before the deadline.

Bench technicians work in well-lit, air-conditioned rooms surrounded by the hum of cooling fans. Avionics equipment manufacturers specify the temperature and humidity limits for testing. While not quite "clean rooms," most have dust collection systems and positive air pressure. Floors, benches, and office furniture are specially made to dissipate static electricity. Screen rooms that block electromagnetic interference are used to test extremely sensitive equipment, or to shield the surrounding areas from powerful radio transmitters.

A Day in the Life of a Avionics Technician

Without avionics technicians, most military and high-tech planes would be unsafe. Avionics technicians test, maintain, and produce aviation electronics, including missile-guidance systems, jet engines, and flight-control circuitry. Much of an avionics job is preventative. Technicians work unusual hours, providing maintenance and support to private research concerns, aerospace companies, the military, and other government agencies. Levels of satisfaction in the industry are high, mainly because it provides intellectual curiosity with a very close attention to detail. The installation of electronics devices, their calibration, and their testing are all critical to the success of any aviation endeavor. Many avionics technicians specialize in one area of expertise, such as microcircuit television microscopy, oscilloscope review, or computerized guidance systems. Since the field is evolving, many people's specialties change over time. Because of rapid changes in technology, continuing education through professional reading, and attending company-sponsored seminars, industry events, and conferences is the norm in this field. Most technicians are also educated through interaction with their colleagues; while each member has an assigned responsibility, the majority of technicians work as part of a team. Communication skills and the ability to write comprehensive and complete reports are as important as technical skills. Two discrete professional categories exist for those in the industry. The first are those involved in the research and development (design and testing stages) of new electronic equipment. These technicians must have curious minds that can imagine potential problems that might occur, including atmospheric conditions, magnetic field interference, and weight limitations. The second category are those involved in the direct installation and maintenance. These people must be extremely attentive to detail, organized, and interested in high degrees of responsibility and long hours. The two fields do cross over at a variety of points, but in general, the fields are separate.

Paying your Dues

Most people attend a specialty school or community college that specializes in electronics engineering for one to three years. Major aerospace employers run their own schools and training centers; but corporate-run schools teach only about each company's own product line. General coursework at these schools includes electronics, the physics of electricity, circuit design, and computer science. Familiarity with math (calculus-level studies are preferred) and a degree of manual dexterity are both helpful. If communications equipment is part of your job, you also will need an Federal Communications Commission (FCC) license as a restricted radio-telephone operator. Most specific skills, such as use of an oscilloscope, or a circuit analyzer, are part of on-the-job training.

Present and Future

Avionics developed with the rise of modern warfare. The number of electronic devices used in navigation, control, maintenance and flight multiplied by a factor of 100 between the World War II B-29 bomber and the current B-58 supersonic bomber. Missile technology, including the development of SCUDS, ICBMs and other "smart" missiles, has also matured. The current demand for aviation technicians is significant. While strong demand exists for technicians now, much of the direction of avionics technicians relies on defense spending, and with budget slashing looming on the horizon, this career faces an uncertain financial future. Aerospace work will continue to be done at some level, but to what degree and by how many people is the question. Research project support and government funding sway from supporting laser-based missile defense systems, to backing propulsion-based systems, from encouraging the development of stealth weapons and high-altitude aircraft, to supporting nothing at all.

Quality of Life

Present and Future

Avionics technicians make the transition from school to the practical work environment during these early, busy years. Many participate in one- to two-month on-the-job training programs sponsored by their aerospace employers. Beginners join research teams as "junior" members and their work is carefully scrutinized by more experienced members. Duties include calibration and installation of communication and other, less complicated, electronics systems. Salaries are reasonable, as are hours, but many entering the field mentioned that the technical journal-reading burden, a feature of most evolving professions, is one unexpected time-stealer.

Five Years out

Five-year survivors are now the senior members of research teams, and, in rare cases, are leading them. Those involved in the design and research stages of products now have significant input into the testing protocols and installation procedures. Those involved in maintenance have regular schedules and may have some oversight responsibility for newer entrants to the field. The majority of avionics technicians--over 70 percent--remain with their original employer through their first five years, possibly due to the paucity of aerospace companies.

Ten Years out

Ten-year avionics technicians spend their time running research projects, working with designers and producers, and managing less-senior avionics technicians. Those who have advanced not only understand the technical specifications and the electronic requirements of what they do, but the spirit of cooperation that must take place for teams in this industry to function efficiently. The majority of ten-year veterans are employed by the major aerospace companies, with the government employing the second largest number. Satisfaction is high and hours increase.

How to Become an Avionics Technician

Essential Information

Avionics technicians repair and maintain the electrical systems in both military and civilian aircrafts. Avionics technicians can opt for on-the-job training and work experience to enter this career field, or they may choose to attend college for a certificate or associate's degree in avionics.

Required Education	No formal education needed for those who complete on-the-job training and work experience; alternately, certificate or associate's degree is required
Other Requirements	A&P certification
Projected Job Growth (2014-2024)*	0% for avionics technicians
Median Salary (2015)*	$58,540 for avionics technicians

Education Requirements for an Avionics Technician

Avionics technicians test and produce various types of aviation electronics, such as jet engines, computerized guidance systems or flight-control circuitry. There are about 170 aviation maintenance technician schools that are certified by the Federal Aviation Administration (FAA). Most of these schools prepare their students for the FAA's airframe and powerplant (A&P) certification.

Curriculum

The curriculum in an avionics degree program includes courses in radio communication theory, avionics system integration and avionics systems line maintenance. Required courses for A&P certification include mathematics, physics, processes and materials, the basics of aircraft, documents and regulations, electricity and power, power plants and fuel and exhaust systems.

While there are undergraduate degrees and certificates available for avionics technicians, graduate and post-graduate degrees in avionics are not commonly offered in these programs. Avionics technicians who are interested in further education can go on to pursue engineering degrees.

On-the-Job Training

Some avionics technicians opt out of formal training and receive on-the-job training. The Federal Aviation Administration requires a minimum 30 months of work experience for A&P certification for non-degreed applicants, during which time technicians must be supervised by licensed avionics professionals. A&P certification applicants must pass an oral, written and practical exam that demonstrates their ability to work in this field. An avionics technician who works on either radio or radar equipment must be licensed by the Federal Communication Commission as a restricted radio-telephone operator.

Career Information

According to the U.S. Bureau of Labor Statistics, there were 17,340 avionics technicians in the United States in 2015. Career opportunities in this field are expected to remain the same (0% growth) from 2014-2024; however, job openings are expected to stem from the high retirement rate among avionics and aircraft repair technicians rather than from industry growth. The average salary for avionics techs was $58,540 per year in 2015, with avionics technicians in scientific research and development services earning the highest median salary.

Avionics technicians are employed at large airports where aircrafts must be serviced and repaired. Repair technicians often diagnose and fix problems with airplanes that have issues on the runway or perform scheduled repairs on aircraft systems. Avionics technicians are also employed at commuter and regional airlines and FAA repair stations.

Avionics technicians need A&P certification in order to work. They can obtain this through an official training or educational program, or they can receive on-the-job training under the supervision of a licensed avionics professional. During training and following certification, avionics technicians help to build, test, and maintain parts for airplanes.

Autopilot Systems

An aircraft automatic pilot system controls the aircraft without the pilot directly maneuvering the controls. The autopilot maintains the aircraft's attitude and/or direction and returns the aircraft to that condition when it is displaced from it. Automatic pilot systems are capable of keeping aircraft stabilized laterally, vertically, and longitudinally.

The primary purpose of an autopilot system is to reduce the work strain and fatigue of controlling the aircraft during long flights. Most autopilots have both manual and automatic modes of operation. In the manual mode, the pilot selects each maneuver and makes small inputs into an autopilot controller. The autopilot system moves the control surfaces of the aircraft to perform the maneuver. In automatic mode, the pilot selects the attitude and direction desired for a flight segment. The autopilot then moves the control surfaces to attain and maintain these parameters.

Autopilot systems provide for one-, two-, or three-axis control of an aircraft. Those that manage the aircraft around only one axis control the ailerons. They are single-axis autopilots, known as wing leveler systems, usually found on light aircraft. Other autopilots are two-axis systems that control the ailerons and elevators. Three-axis autopilots control the ailerons, elevators, and the rudder. Two-and three-axis autopilot systems can be found on aircraft of all sizes.

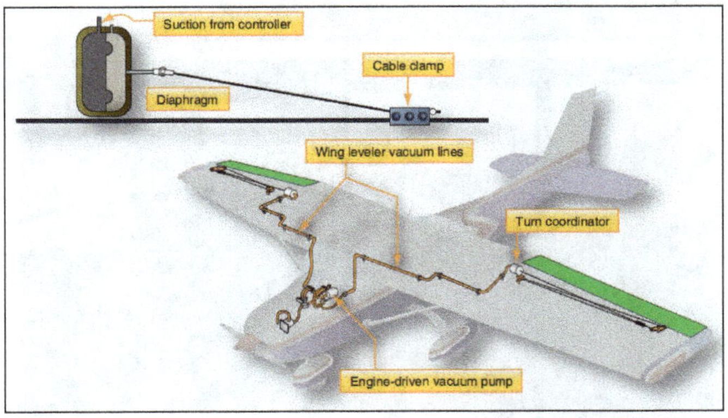

The wing leveler system on a small aircraft is a vacuum-operated single-axis autopilot. Only the ailerons are controlled. The aircraft's turn coordinator is the sensing element. Vacuum from the instrument vacuum system is metered to the diaphragm cable actuators to move the ailerons when the turn coordinator senses roll.

There are many autopilot systems available. They feature a wide range of capabilities and complexity. Light aircraft typically have autopilots with fewer capabilities than highperformance and transport category aircraft. Integration of navigation functions is common, even on light aircraft autopilots. As autopilots increase in complexity, they not only manipulate the flight control surfaces, but other flight parameters as well.

Some modern small aircraft, high-performance, and transport category aircraft have very elaborate autopilot systems known as automatic flight control systems (AFCS). These three-axis systems go far beyond steering the airplane. They control the aircraft during climbs, descents, cruise, and approach to landing. Some even integrate an auto-throttle function that automatically controls engine thrust that makes autolandings possible.

For further automatic control, flight management systems have been developed. Through the use of computers, an entire flight profile can be programmed ahead of time allowing the pilot to supervise its execution. An FMS computer coordinates nearly every aspect of a flight, including the autopilot and auto throttle systems, navigation route selection, fuel management schemes, and more.

Basis for Autopilot Operation

The basis for autopilot system operation is error correction. When an aircraft fails to meet the conditions selected, an error is said to have occurred. The autopilot system automatically corrects that error and restores the aircraft to the flight attitude desired by the pilot. There are two basic ways modern autopilot systems do this. One is position based and the other is rate based. A position based autopilot manipulates the aircraft's controls so that any deviation from the desired attitude of the aircraft is corrected. This is done by memorizing the desired aircraft attitude and moving the control surfaces so that the aircraft returns to that attitude. Rate based autopilots use information about the rate of movement of the aircraft, and move control surfaces to counter the rate of change that causes the error. Most large aircraft use rate-based autopilot systems. Small aircraft may use either.

How to Maintain Autopilot System

Maintenance of an autopilot system consists of visual inspection, replacement of components, cleaning, lubrication, and an operational checkout of the system. Consult the manufacturer's maintenance manual for all of these procedures.

Step 1

With the autopilot disengaged, the flight controls should function smoothly. The resistance offered by the autopilot servos should not affect the control of the aircraft. The interconnecting mechanisms between the autopilot system and the flight control system should be correctly aligned and smooth in operation. When applicable, the operating cables should be checked for tension.

Step 2

An operational check is important to assure that every circuit is functioning properly. An autopilot operational check should be performed on new installations, after replacement of an autopilot component, or whenever a malfunction in the autopilot is suspected.

Step 3

After the aircraft's main power switch has been turned on, allow the gyros to come up to speed and the amplifier to warm up before engaging the autopilot. Some systems are designed with safeguards that prevent premature autopilot engagement. While holding the control column in the normal flight position, engage the autopilot system using the switch on the autopilot controller.

After the system is engaged, perform the operational checks specified for the particular aircraft. In general, the checks are as follows:

1. Rotate the turn knob to the left; the left rudder pedal should move forward, and the control column wheel should move to the left and slightly aft.
2. Rotate the turn knob to the right; the right rudder pedal should move forward, and the control column wheel should move to the right and slightly aft. Return the turn knob to the center position; the flight controls should return to the level-flight position.
3. Rotate the pitch-trim knob forward; the control column should move forward.
4. Rotate the pitch-trim knob aft; the control column should move aft.

Step 4

If the aircraft has a pitch-trim system installed, it should function to add down-trim as the control column moves forward and add up-trim as the column moves aft. Many pitch-trim systems have an automatic and a manual mode of operation. The above action occurs only in the automatic mode.

Step 5

Check to see if it is possible to manually override or overpower the autopilot system in all control positions. Center all the controls when the operational checks have been completed.

Disengage the autopilot system and check for freedom of the control surfaces by moving the control columns and rudder pedals. Then, reengage the system and check the emergency disconnect release circuit. The autopilot should disengage each time the release button on the control yoke is actuated.

When performing maintenance and operational checks on a specific autopilot system, always follow the procedure recommended by the aircraft or equipment manufacturer.

Radio Communication

Much of aviation communication and navigation is accomplished through the use of radio waves. Communication by radio was the first use of radio frequency transmissions in aviation.

Radio Waves

A radio wave is invisible to the human eye. It is electromagnetic in nature and part of the electronic spectrum of wave activity that includes gamma rays, x-rays, ultraviolet rays, infrared waves, and visible light rays, as well all radio waves. The atmosphere is filled with these waves. Each wave occurs at a specific frequency and has a corresponding wavelength. The relationship between frequency and wavelength is inversely proportional. A high frequency wave has a short wave length and a low frequency wave has a long wave length.

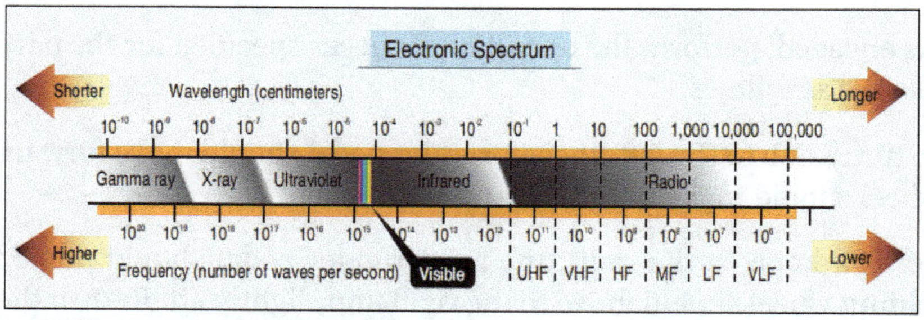

Radio waves are just some of the electromagnetic waves found in space.

In aviation, a variety of radio waves are used for communication. illustrates the radio spectrum that includes the range of common aviation radio frequencies and their applications.

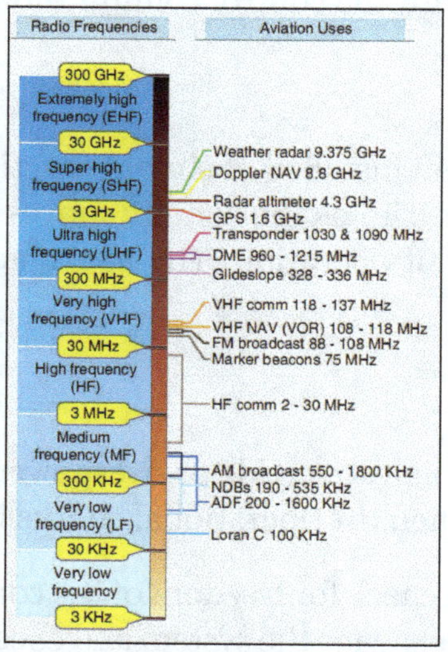

There is a wide range of radio frequencies. Only the very low frequencies and the extremely high frequencies are not used in aviation.

A wide range of frequencies are used from low frequency (LF) at 100 kHz (100,000 cycles per second) to super high frequency (SHF) at nearly 10gHz (10,000,000,000 cycles per second). The Federal Communications Commission (FCC) controls the assignment of frequency usage.

AC power of a particular frequency has a characteristic length of conductor that is resonant at that frequency. This length is the wavelength of the frequency that can be seen on an oscilloscope. Fractions of the wavelength also resonate, especially half of a wavelength, which is the same as half of the AC sign wave or cycle.

The frequency of an AC signal is the number of times the AC cycles every second. AC applied to the center of a radio antenna, a conductor half the wavelength of the AC frequency, travels the length of the antenna, collapses, and travels the length of the antenna in the opposite direction. The number of times it does this every second is known as the radio wave signal frequency or radio frequency as shown in figure. As the current flows through the antenna, corresponding electromagnetic and electric fields build, collapse, build in the opposite direction, and collapse again.

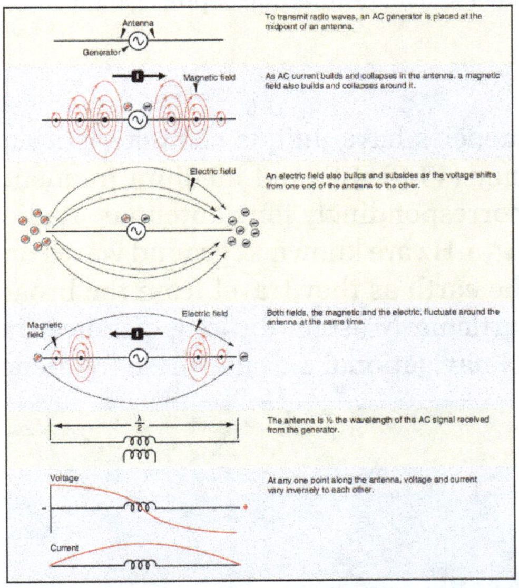

Radio waves are produced by applying an AC signal to an antenna. This creates a magnetic and electric field around the antenna. They build and collapse as the AC cycles. The speed at which the AC cycles does not allow the fields to completely collapse before the next fields build. The collapsing fields are then forced out into space as radio waves.

To transmit radio waves, an AC generator is placed at the midpoint of an antenna. As AC current builds and collapses in the antenna, a magnetic field also builds and collapses around it. An electric field also builds and subsides as the voltage shifts from one end of the antenna to the other. Both fields, the magnetic and the electric, fluctuate around the antenna at the same time. The antenna is half the wavelength of the AC signal received from the generator. At any one point along the antenna, voltage and current vary inversely to each other.

Because of the speed of the AC, the electromagnetic fields and electric fields created around the antenna do not have time to completely collapse as the AC cycles. Each new current flow creates new fields around the antenna that force the nottotally- collapsed fields from the previous AC cycle out into space. These are the radio waves. The process is continuous as long as AC is applied to the antenna. Thus, steady radio waves of a frequency determined by the input AC frequency propagate out into space.

Radio waves are directional and propagate out into space at 186,000 miles per second. The distance they travel depends on the frequency and the amplification of the signal AC sent to the antenna. The electric field component and the electromagnetic field component are oriented at 90° to each other, and at 90° to the direction that the wave is traveling.

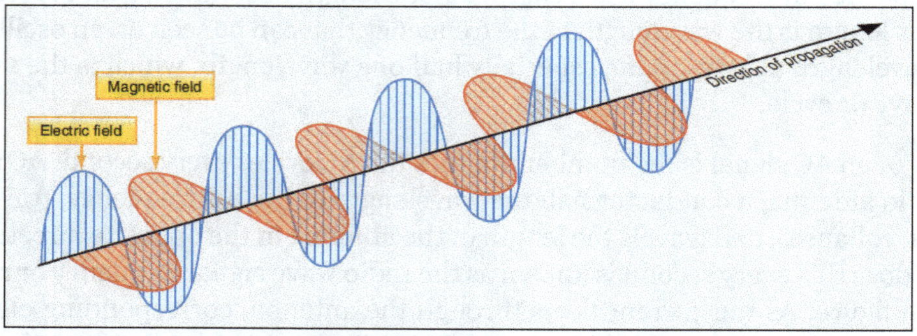

The electric field and the magnetic field of a radio wave are perpendicular to each other and to the direction of propagation of the wave.

Types of Radio Waves

Radio waves of different frequencies have unique characteristics as they propagate through the atmosphere. Very low frequency (VLF), LF, and medium frequency (MF) waves have relatively long wavelengths and utilize correspondingly long antennas. Radio waves produced at these frequencies ranging from 3kHz to 3mHz are known as ground waves or surface waves. This is because they follow the curvature of the earth as they travel from the broadcast antenna to the receiving antenna. Ground waves are particularly useful for long distance transmissions. Automatic direction finders (ADF) and LORAN navigational aids use these frequencies.

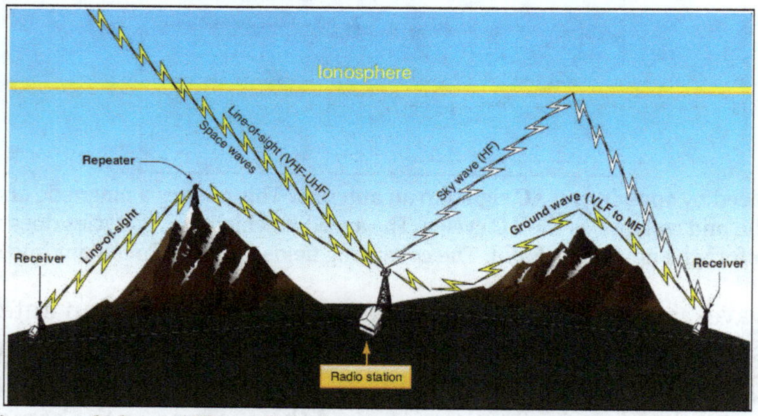

Radio waves behave differently in the atmosphere depending in their frequency.

High frequency (HF) radio waves travel in a straight line and do not curve to follow the earth's surface. This would limit transmissions from the broadcast antenna to receiving antennas only in the line-of-sight of the broadcast antenna except for a unique characteristic. HF radio waves bounce off of the ionosphere layer of the atmosphere. This refraction extends the range of HF signals beyond line-of-sight. As a result, transoceanic aircraft often use HF radios for voice communication. The frequency range is between 2 to 25 MHz. These kinds of radio waves are known as sky waves.

Above HF transmissions, radio waves are known as space waves. They are only capable of line-of-sight transmission and do not refract off of the ionosphere. Most aviation communication and navigational aids operate with space waves. This includes VHF (30-300MHz), UHF (300MHz-3GHz), and super high frequency (SHF) (3Ghz- 30Ghz) radio waves.

VHF communication radios are the primary communication radios used in aviation. They operate in the frequency range from 118.0 MHz to 136.975MHz. Seven hundred and twenty separate and distinct channels have been designated in this range with 25 kilohertz spacing between each channel. Further division of the bandwidth is possible, such as in Europe where 8.33 kilohertz separate each VHF communication channel. VHF radios are used for communications between aircraft and air traffic control (ATC), as well as air-to-air communication between aircraft. When using VHF, each party transmits and receives on the same channel. Only one party can transmit at any one time.

Antennas

As stated, antennas are conductors that are used to transmit and receive radio frequency waves. Although the airframe technician has limited duties in relation to maintaining and repairing avionics, it is the responsibility of the technician to install, inspect, repair, and maintain aircraft radio antennas.

Three characteristics are of major concern when considering antennas:

1. Length
2. Polarization
3. Directivity

The exact shape and material from which an antenna is made can alter its transmitting and receiving characteristics. Also note that some non-metallic aircraft have antennas imbedded into the composite material as it is built up.

Length

When an AC signal is applied to an antenna, it has a certain frequency. There is a corresponding wavelength for that frequency. An antenna that is half the length of this wavelength is resonant. During each phase of the applied AC, all voltage and current values experience the full range of their variability. As a result, an antenna that is half the wavelength of the corresponding AC frequency is able to allow full voltage and full current flow for the positive phase of the AC signal in one direction. The negative phase of the full AC sign wave is accommodated by the voltage and current simply changing direction in the conductor. Thus, the applied AC frequency flows through its entire wavelength, first in one direction and then in the other. This produces the strongest signal to be radiated by the transmitting antenna. It also facilitates capture of the wave and maximum induced voltage in the receiving antenna.

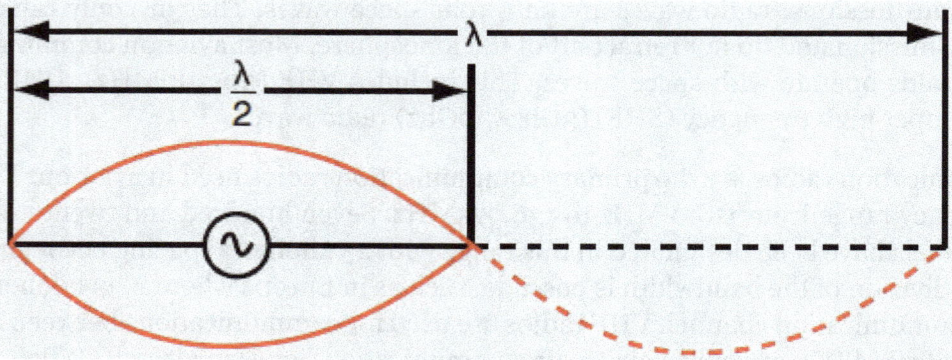

An antenna equal to the full length of the applied AC frequency wavelength would have the negative cycle current flow along the antenna as shown by the dotted line. An antenna that is ½ wavelength allows current to reverse its direction in the antenna during the negative cycle. This results in low current at the ends of the ½ wavelength antenna and high current in the center. As energy radiates into space, the field is strongest 90° to the antenna where the current flow is strongest.

Most radios, especially communication radios, use the same antenna for transmitting and receiving. Multichannel radios could use a different length antenna for each frequency, however, this is impractical. Acceptable performance can exist from a single antenna half the wavelength of a median frequency. This antenna can be made effectively shorter by placing a properly rated capacitor in series with the transmission line from the transmitter or receiver. This electrically shortens the resonant circuit of which the antenna is a part. An antenna may be electrically lengthened by adding an inductor in the circuit. Adjusting antenna length in this fashion allows the use of a single antenna for multiple frequencies in a narrow frequency range.

Many radios use a tuning circuit to adjust the effective length of the antenna to match the wavelength of the desired frequency. It contains a variable capacitor and an inductor connected in parallel in a circuit. Newer radios use a more efficient tuning circuit. It uses switches to combine frequencies from crystal controlled circuits to create a resonant frequency that matches the desired frequency. Either way, the physical antenna length is a compromise when using a multichannel communication or navigation device that must be electronically tuned for the best performance.

A formula can be used to find the ideal length of a half wavelength antenna required for a particular frequency as follows:

$$\text{Antenna Length (feet)} = \frac{468}{\text{F MHz}}$$

The formula is derived from the speed of propagation of radio waves, which is approximately 300 million meters per second. It takes into account the dielectric effect of the air at the end of an antenna that effectively shortens the length of the conductor required.

VHF radio frequencies used by aircraft communication radios are 118–136.975 MHz. The corresponding half wavelengths of these frequencies are 3.96 – 3.44 feet (47.5–41.2 inches). Therefore,

VHF antennas are relatively long. Antennas one-quarter of the wavelength of the transmitted frequency are often used. This is possible because when mounted on a metal fuselage, a ground plane is formed and the fuselage acts as the missing one-quarter length of the half wavelength antenna.

Polarization, Directivity and Field Pattern

Antennas are polarized. They radiate and receive in certain patterns and directions. The electric field cause by the voltage in the conductor is parallel to the polarization of an antenna. It is caused by the voltage difference between each end of the antenna. The electromagnetic field component of the radio wave is at 90° to the polarization. It is caused by changing current flow in the antenna. As radio waves radiate out from the antenna they propagate in a specific direction and in a specific pattern. This is the antenna field. The orientation of the electric and electromagnetic fields remains at 90° to each other, but radiate from antenna with varying strength in different directions. The strength of the radiated field varies depending on the type of antenna and the angular proximity to it. All antennas, even those that are omnidirectional, radiate a stronger signal in some direction compared to other directions. This is known as the antenna field directivity.

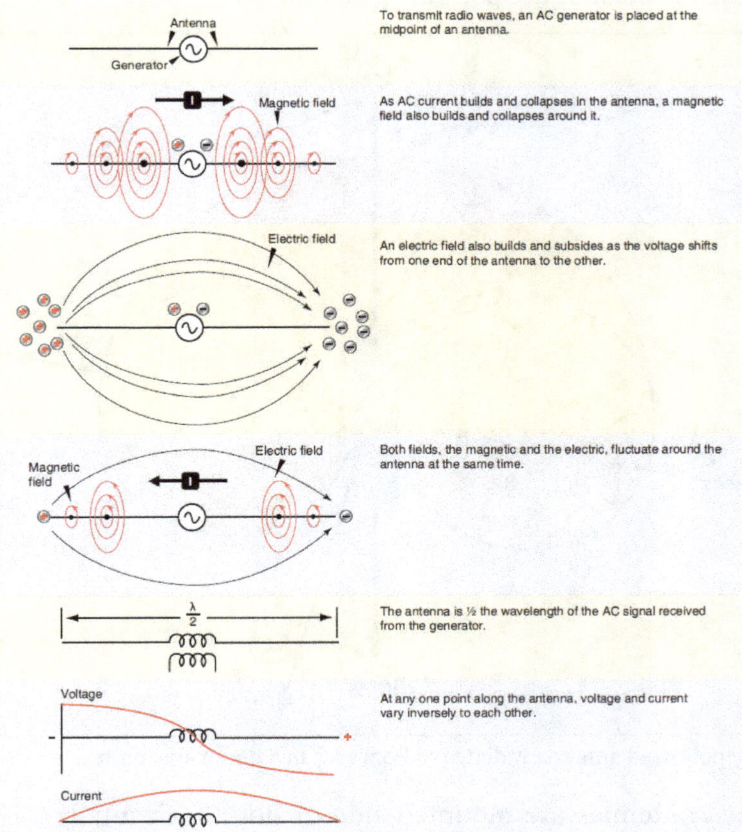

Radio waves are produced by applying an AC signal to an antenna. This creates a magnetic and electric field around the antenna. They build and collapse as the AC cycles. The speed at which the AC cycles does not allow the fields to completely collapse before the next fields build. The collapsing fields are then forced out into space as radio waves.

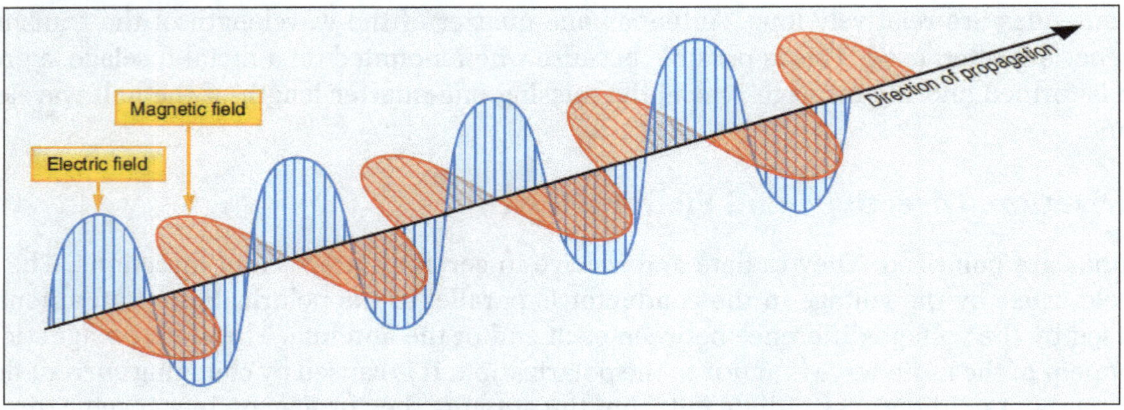

The electric field and the magnetic field of a radio wave are perpendicular to each other and to the direction of propagation of the wave.

Receiving antennas with the same polarization as the transmitting antenna generate the strongest signal. A vertically polarized antenna is mounted up and down. It radiates waves out from it in all directions. To receive the strongest signal from these waves, the receiving antenna should also be positioned vertically so the electromagnetic component of the radio wave can cross it at as close to a 90° angle as possible for most of the possible proximities.

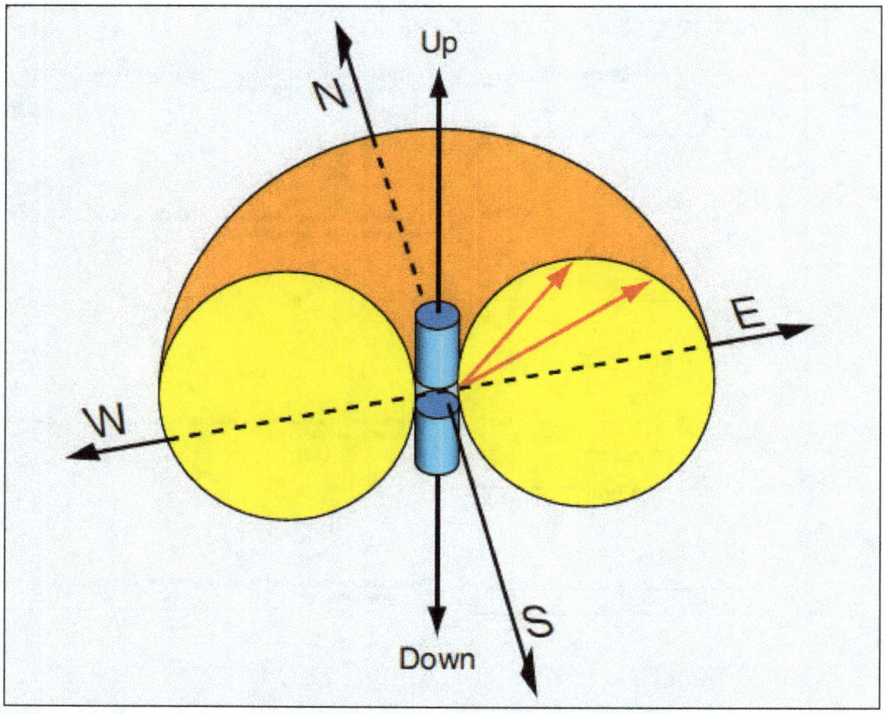

A vertically polarized antenna radiates radio waves in a donut-like pattern in all directions.

Horizontally polarized antennas are mounted side to side (horizontally). They radiate in a donut-like field. The strongest signals come from, or are received at, 90° to the length of the antenna. There is no field generated off of the end of the antenna. illustrates the field produced by a horizontally polarized antenna.

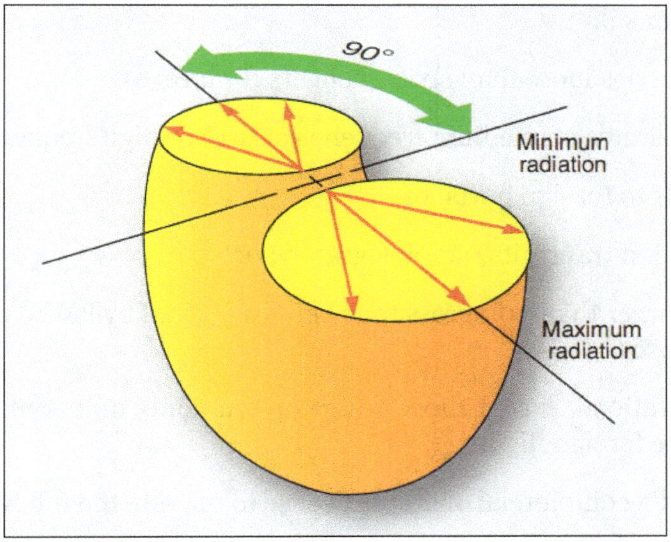

A horizontally polarized antenna radiates in a donut-like pattern.
The strongest signal is at 90° to the length of the conductor.

Many vertical and horizontal antennas on aircraft are mounted at a slight angle off plane. This allows the antenna to receive a weak signal rather than no signal at all when the polarization of the receiving antenna is not identical to the transmitting antenna.

How to get Licenced as a Commercial Radio Operator

FCC rules require that licensees of ship and aircraft stations permit only persons holding the appropriate FCC-issued commercial operator license to perform specified transmitter operation, maintenance and repair duties. To obtain a new or upgraded FCC commercial operator license, you must pass an examination administered by one of the Commercial Operator License Examination Managers (COLEMs).

Who Needs a Commercial Operator License

You need a commercial operator license to operate, and/or to repair and maintain, specified ship, and aircraft radio communication stations. The licensing requirements for operating such radio stations and the licensing requirements for repairing and maintaining such radio stations are discussed separately below.

CAUTION: These listings only describe when a commercial operator license is necessary. A commercial operator license does not constitute or imply FCC authorization to transmit radio signals. Before you operate any radio station, make certain that the station is licensed as required by the FCC.

Radio Operations

You need a commercial radio operator license to operate the following:

- Ship radio stations if:
  1. the vessel carries more than six passengers for hire; or
  2. the radio operates on medium frequencies (MF) or high frequencies (HF); or
  3. the ship sails to foreign ports; or
  4. the ship station transmits radiotelegraphy; or
  5. the ship is larger than 300 gross tons and is required by law to carry a radio station for safety purposes.
- Aircraft radio stations, except those which operate only on very high frequencies (VHF) and do not make foreign flights.

You do sort by not need a commercial operator license to operate the following:

- Ship stations operating only on VHF frequencies that do not travel to foreign ports or make international communications (unless the vessel carries more than six passengers for hire, or the ship is larger than 300 gross tons and is required by law to carry a radio station for safety purposes).
- Shore radar, shore radiolocation, maritime support or shore radionavigation stations.
- Survival craft stations or EPIRBs.
- Ship radar stations, if (a) the radar frequency is determined by a nontunable, pulse type magnetron or other fixed tune device, and (b) the radar is capable of being operated exclusively by external controls.
- Coast stations.
- Aircraft stations which operate only on VHF frequencies and do not make foreign flights.
- Aircraft radar sets, radio altimeters, transponders or other aircraft automatic radionavigation transmitters.
- ELTs or aviation survival craft stations used solely for survival purposes.

Radio Maintenance and Repair

You need a commercial operator license to repair and maintain the following:

- All ship radio and radar stations.
- All coast stations.
- All hand-carried units used to communicate with ships and coast stations on marine frequencies.
- All aircraft stations and aeronautical ground stations (including hand-carried portable units) used to communicate with aircraft.

You do sort by not need a commercial radio operator license to operate, repair, or maintain any of the following types of stations:

- Two-way land mobile radio equipment, such as that used by police and fire departments, taxicabs and truckers, businesses and industries, ambulances and rescue squads, and local, state, and federal government agencies.
- Personal radio equipment used in the Citizens Band Radio Service (CBRS), Radio Control Radio Service (RCRS), and General Mobile Radio Services (GMRS).
- Auxiliary broadcast stations, such as remote pickup stations.
- Domestic public fixed and mobile radio systems, such as mobile telephone systems, cellular systems, rural radio systems, point-to-point microwave systems, multipoint distribution systems, etc.
- Stations that operate in the Cable Television Relay Service.
- Satellite stations, both uplink and downlink of all types.

Additional Licensee Information

Operating While Application is Pending

A person who has filed an application for a commercial operator license with the FCC, and who holds a Proof of Passing Certificate(s) indicating that he or she has passed the necessary examination(s) within the previous 365 days, is authorized to exercise the rights and privileges of the applied-for operator license during the period before the FCC acts on the application, but only for a period of sort by 90 days from the date the application was filed, sort by provided that the applicant has not had a commercial radio operator license revoked or suspended in the past, and is not the subject of an ongoing suspension proceeding.

Renewal Procedures

If you need to renew a commercial radio operator license, submit completed FCC Forms 159 (Fee Processing Form) and 605. A fee is required. Current information on the fee amount and filing location can be found in the most recent edition of the Wireless Telecommunications Bureau Fee Filing Guide, FCC Form 1070Y, or by calling the Customer Support Hotline at (877) 480-3201 (select option 2 after the main menu recording). The WTB Fee Filing Guide and FCC Form 1070Y are also available from the fax-on-demand system by dialing (202) 418-0177 from the handset of a fax machine.

If you file an application to renew your license before it expires, you may continue to operate under the authority of your license while the FCC processes your renewal application. However, if you fail to renew your license before it expires, you cannot operate or maintain equipment that requires that license until it is renewed. You may file to renew your expired license any time during the five-year grace period after your license expires. If you fail to renew your license within the grace period, you must apply for a new license and re-take the required examination(s).

Posting of License or Permit

If you are employed at a station where your operator license must be posted, you may post either your proof of passing certificate or a signed copy of your renewal application instead of your license while your application for license renewal is being processed.

Lost, Stolen, Mutilated or Destroyed Licenses

You may apply for a duplicate license by submitting FCC Forms 159 (Fee Processing Form) and 605 according to the instructions printed on the 605 application. A fee is required. Current information on the fee amount and filing location can be found in the most recent edition of the Wireless Telecommunications Bureau Fee Filing Guide, FCC Form 1070Y, or by calling the FCC's Consumer Center toll free at 1-888-CALL-FCC (225-5322). The WTB Fee Filing Guide and FCC Form 1070Y are also available from the fax-on-demand system by dialing (202) 418-0177 from the handset of a fax machine. In the alternative, a licensee can print a new license document by following the procedures for providing access to official electronic authorizations.

Name Changes

If you change your name, you must apply for a replacement license in your new legal name. Be sure to indicate the reason for your application and give both your former and new legal names.

How to Pre-Flight an Aircraft

Preflighting an airplane is the process of inspecting it prior to takeoff. This is a general outline of steps to assure the airworthiness of your aircraft for takeoff. Actual preflight requirements may vary from one type aircraft to another, and these steps would not be appropriate for military or commercial aircraft.

Steps

1. Use a preflight checklist if you have one. Many rental or fleet aircraft have one which may have all recent service, inspection, and maintenance information updated for each aircraft. Here are some things you may see:

- Flight hours. Because aircraft maintenance is usually scheduled at flight hour intervals, the actual hours flown are logged to accommodate scheduling service dates for these activities.
- Pilot observations. When more than one pilot is likely to operate an aircraft, it is beneficial to have input from each pilot as to the flight characteristics of the plane. One pilot may notice a vibration, shudder, or other physical anomaly, or unusual readouts from gauges while in flight, that the next pilot should be aware of.
- Service schedule. If a component (airframe, engine, etc) has a scheduled 100 hrs inspection

due after 5 more flight hours, a longer flight could be postponed until the service is performed, or another aircraft utilized.

2. Begin your preflight in the cockpit, since before flying, you will need to make sure all cockpit equipment is operating, and the fuel tanks have sufficient fuel for your flight.

- Make sure the aircraft registration, certification, and other paperwork is in the cabin and up to date.
- Make sure the ignition switch is in the off position.
- Turn on the master power switch.
- Check the fuel gauges. If the fuel is low, call for the service truck to bring your fuel out while you continue to do the rest of your checks.
- Listen to the sounds of equipment powering on. Radio cooling fans, instrument gyros, and other equipment make noticeable sounds, and when unusual sounds are heard, it may be a warning an instrument or radio could fail in-flight.
- Check flaps, landing gear lockdown levers, and other flight controllers for smooth, normal function.

3. Exit the aircraft.

- As you climb out of the cabin, take a look at the seat support rails (on small aircraft) to make sure the seat is anchored securely and all fasteners are in place.
- Check the cabin door to be sure it operates smoothly and shuts securely. Worn hinges or latches that do not secure properly can cause an inflight emergency. If the door does not move smoothly into open and closed positions, it may indicate the airframe and internal structure are damaged.

4. Walk around the aircraft, looking for damage cause by impacts or cracks and seams separating from airframe fatigue, hard landings, or other mishaps.

5. Begin at the right wing, after exiting the aircraft, and look at the forward wing surface for nicks,

loose fasteners, dents, or other damage. Look at the flight control surfaces, flaps, and ailerons. Make sure that items are secure and have no loose fasteners.

6. Remove the fuel cap (for wing fuel tanks) and visually confirm they have sufficient fuel to make your flight. Replace the fuel cap securely.

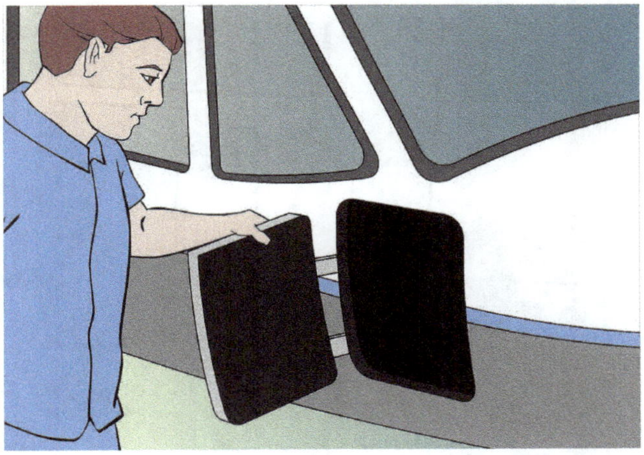

7. Check stall indicator (depending on which wing it is mounted on), struts, if applicable, and other features on the right wing.

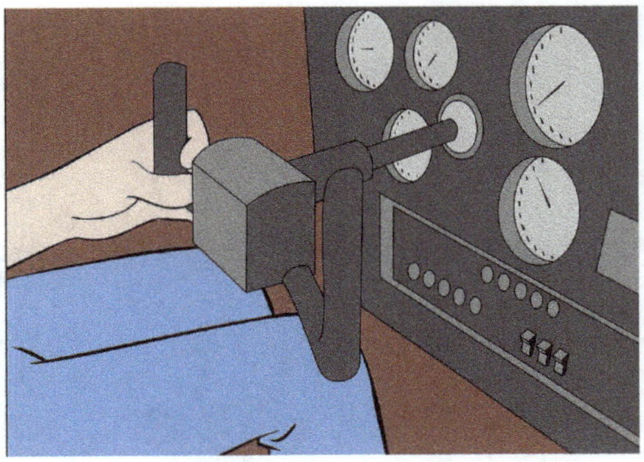

8. Move down the right side of the aircraft. Continue looking at the aircraft surface. Be particularly

careful to observe defects or loose fasteners in the aircraft surface.

9. Move to the tail assembly. While at the tail, you may remove the wheel chock or tail tie-down. Look at the elevators and the rudder. As with all control surfaces, these should be tight, with no loose motion or free play.

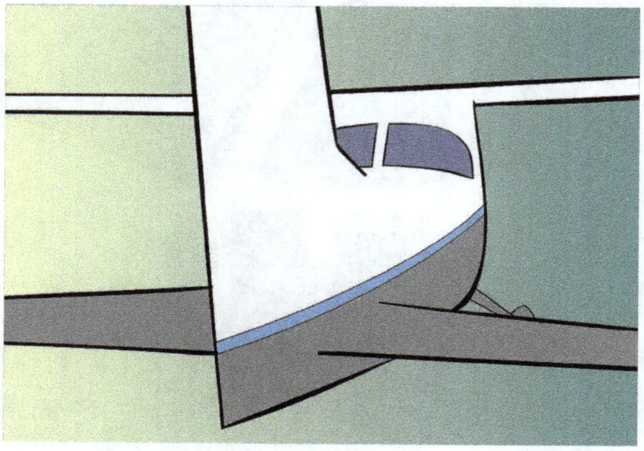

10. Look at the antennae assembly, which is usually located at the tail, as well as the tail wheel, to be sure no lubricants or brake fluid is leaking, and that the tire is properly inflated. Give the gear suspension a once-over, too, making sure any boots or covers are in place, and that all support cables are tight.

11. Move to the opposite side of the aircraft, continuing to look over the aircraft skin to the wing. Remove the fuel cap and look into the tank to be sure it is fueled, replace the tank cap securely, and check out the control surfaces (again, the ailerons and flaps).

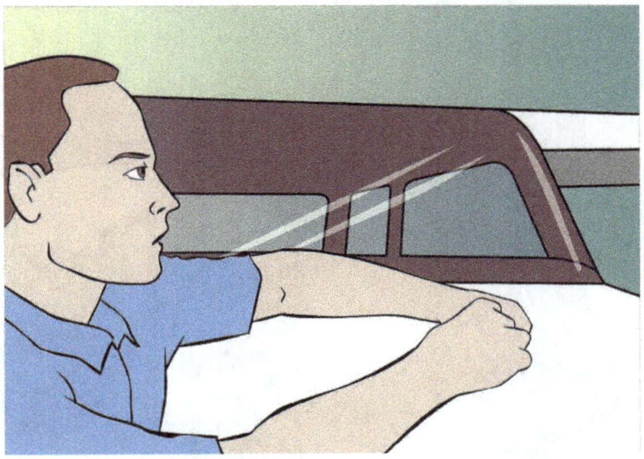

12. Move to the front of the aircraft and look at the exhaust; looking for oil blow-out and other damage.

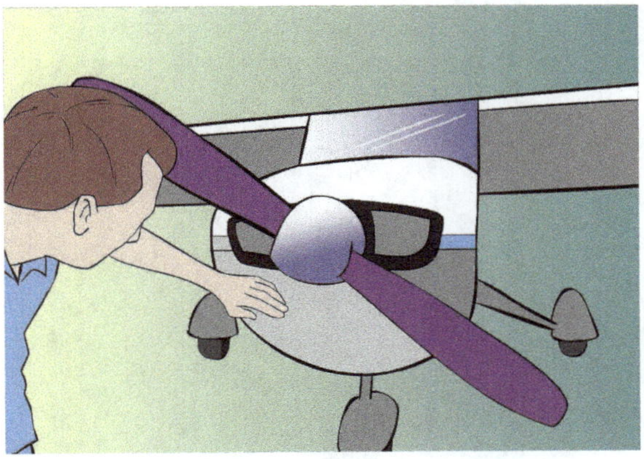

13. Check the engine oil, the ignition wires, magneto electrical connections, and the fuel lines and other hoses to be sure they are seated properly and tightly clamped. Look at the alternator belt to make sure it is tight, and the air intake to be sure it is not obstructed.

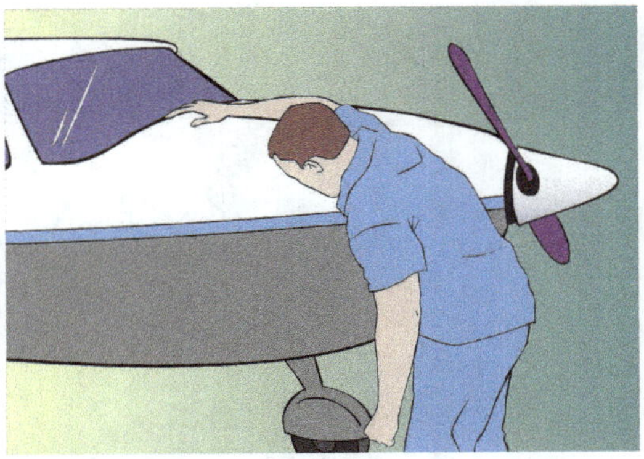

14. Move to the propeller. Never place your body in the swing radius of an airplane propeller.

Check the "spinner" for lubricant leaks, missing bolts and pins, or other problems. Look at the propeller itself, to be sure the blades are not cracked, bent, delaminated, or damaged in other ways.

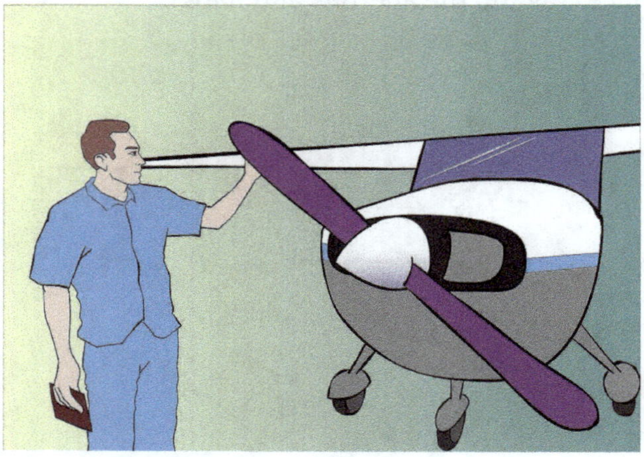

15. Check for fuel or lubricant leaks around the engine compartment, cowling, and fuel tank locations. Any leaks or appearance of fluid on the aircraft skin should be inspected by a trained maintenance person prior to flight.

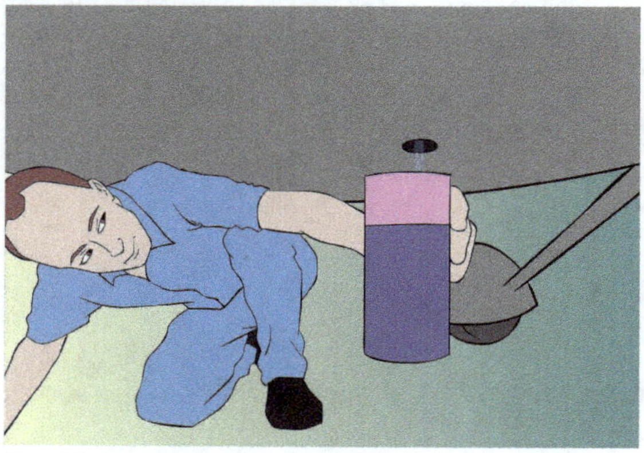

16. Take a look at the wheels, landing gear, and landing gear compartment doors. Look for loose fittings, tires which may have tread separation, low inflation pressure, and other defects.

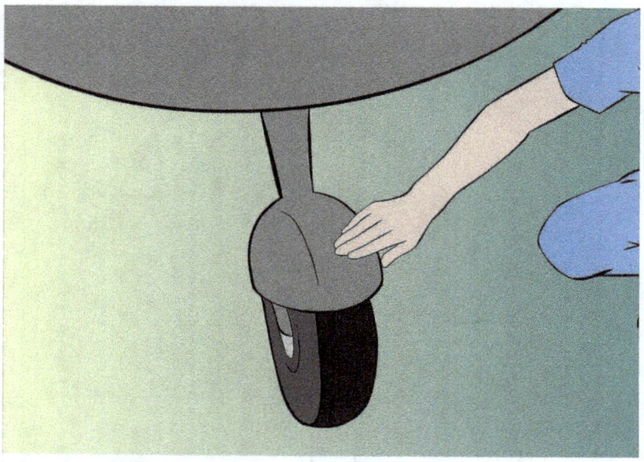

17. Remove the wing tie downs and wheel chocks, walk to the front of the aircraft, and take a long, careful look to make sure you have not overlooked anything.

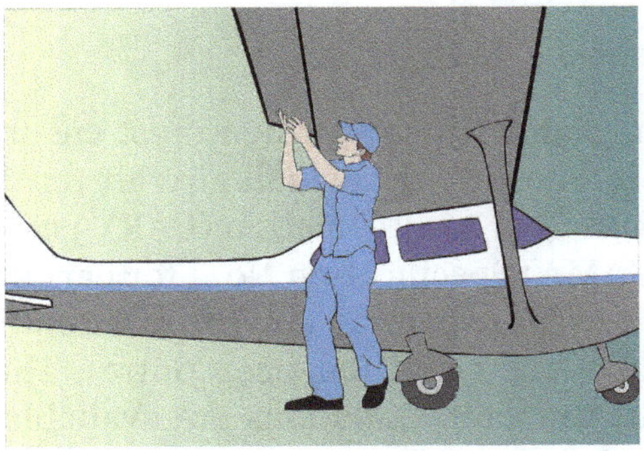

Weight and Balance Control

Aircraft can perform safely and achieve their designed efficiency only when they are operated and maintained in the way their designers intended. This safety and efficiency is determined to a large degree by holding the aircraft's weight and balance parameters within the limits specifiedfor its design. The remainder of this handbook describes how this is done.

Responsibility for Weight and Balance Control

The responsibility for proper weight and balance control begins with the engineers and designers and extends to the technicians who maintain the aircraft and the pilots who operate them. Modern aircraft are engineered utilizing stateof-the-art technology and materials to achieve maximum reliability and performance for the intended category. As much care and expertise must be exercised in operating and maintaining these efficient aircraft as was taken in their design and manufacturing:

1. The designers of an aircraft set the maximum weight based on the amount of lift the wings or rotors can provide under the operational conditions for which the aircraft is designed. The structural strength of the aircraft also limits the maximum weight the aircraft can safely carry. The designers carefully determine the ideal center of gravity (CG) and calculate the maximum allowable deviation from this specific location.
2. The manufacturer provides the aircraft operator with the empty weight of the aircraft and the location of its empty weight center of gravity (EWCG) at the time the certified aircraft leaves the factory. Amateur-built aircraft must have this information determined and available at the time of certification
3. The FAA-certificated mechanic or repairman who maintains the aircraft keeps the weight and balance records current, recording any changes that have been made because of repairs or alterations.
4. The pilot in command (PIC) has the responsibility prior to every flight to know the maximum allowable weight of the aircraft and its CG limits. This allows the pilot to determine

during the preflight inspection that the aircraft is loaded so that the CG is within the allowable limits.

Terminology

Pilots and FAA-certificated mechanics or repairmen must ensure they understand the terms as they relate to the aircraft in question. For small aircraft terminology, use the information found in sources associated with Civil Air Regulation (CAR) 3 certification or General Aviation Manufacturers Association (GAMA) Specification No. 1 for part 23 aircraft or part 27 for rotorcraft. For terminology applied to large part 25 aircraft, information can be found in Advisory Circular (AC) 120-27, Aircraft Weight and Balance Control. The glossary contains the most current terms and defin tions. Current regulations are available from the Superintendent of Documents; U.S. Government Printing Office; Washington, DC 20402. Earlier regulations may be available in libraries or in the Federal Register.

Weight Control

Weight is a major factor in airplane construction and operation, and it demands respect from all pilots and particular diligence by all maintenance personnel. Excessive weight reduces the efficiency of an aircraft and the available safety margin if an emergency condition should arise.

When an aircraft is designed, it is made as light as the required structural strength allows, and the wings or rotors are designed to support the maximum allowable weight. When the weight of an aircraft is increased, the wings or rotors must produce additional lift and the structure must support not only the additional static loads, but also the dynamic loads imposed by flight maneuvers. For example, the wings of a 3,000-pound airplane must support 3,000 pounds in level flight, but when the airplane is turned smoothly and sharply using a bank angle of 60°, the dynamic load requires the wings to support twice this or 6,000 pounds.

Severe uncoordinated maneuvers or flight into turbulence can impose dynamic loads on the structure great enough to cause failure. In accordance with Title 14 of the Code of Federal Regulations (14 CFR) part 23, the structure of a normal category airplane must be strong enough to sustain a load factor of 3.8 times its weight. Every pound of weight added to a normal category aircraft requires that the structure be strong enough to support 3.8 pounds. An aircraft operated in the utility category must sustain a load factor of 4.4 times its weight, and acrobatic category aircraft must be strong enough to withstand 6.0 times their weight.

The lift produced by a wing is determined by its airfoil shape, angle of attack, speed through the air, and air density. When an aircraft takes off from an airport with a high density altitude, it must accelerate to a speed faster than would be required at sea level to produce enough lift to allow takeoff; therefore, a longer takeoff run is necessary. The distance needed may be longer than the available runway. When operating from a high density altitude airport, the Pilot's Operating Handbook (POH) or Airplane Flight Manual (AFM) must be consulted to determine the maximum weight allowed for the aircraft under the conditions of altitude, temperature, wind, and runway conditions.

Effects of Weight

Most modern aircraft are so designed that, when all seats are occupied, the baggage compartment is full, and all fuel tanks are full, the aircraft is grossly overloaded. This type of design requires the pilot to give great consideration to the requirements of each specific flight. If maximum range is required, occupants or baggage must be left behind, or if the maximum load must be carried, the range, dictated by the amount of fuel on board, must be reduced.

Overloading an aircraft can create a variety of problems:

- The aircraft needs a higher takeoff speed, which results in a longer takeoff run.
- Both the rate and angle of climb are reduced.
- The service ceiling is lowered.
- The cruising speed is reduced.
- The cruising range is shortened.
- Maneuverability is decreased.
- A longer landing roll is required because the landing speed is higher.
- Excessive loads are imposed on the structure, especially the landing gear.

The POH or AFM includes tables or charts that give the pilot an indication of the performance expected for any weight. An important part of careful preflight planning includes a check of these charts to determine if the aircraft is loaded so the proposed flight can be safely made.

Weight Changes

The maximum allowable weight for an aircraft is determined by design considerations. However, the maximum operational weight may be less than the maximum allowable weight due to such considerations as high density altitude or high-drag field conditions caused by wet grass or water on the runway. The maximum operational weight may also be limited by the departure or arrival airport's runway length.

One important preflight consideration is the distribution of the load in the aircraft. Loading the aircraft so the gross weight is less than the maximum allowable is not enough. This weight must be distributed to keep the CG within the limits specified in the POH or AFM.

If the CG is too far forward, a heavy passenger can be moved to one of the rear seats or baggage may be shifted from a forward baggage compartment to a rear compartment. If the CG is too far aft, passenger weight or baggage can be shifted forward. The fuel load should be balanced laterally. The pilot should pay special attention to the POH or AFM regarding the operation of the fuel system in order to keep the aircraft balanced in flight.

Weight and balance of a helicopter is far more critical than for an airplane. Some helicopters may be properly loaded for takeoff, but near the end of a long flight when the fuel tanks are almost empty, the CG may have shifted enough for the helicopter to be out of balance laterally or longitudinally. Before making any long flight, the CG with the fuel available for landing must be checked to ensure it is within the allowable range.

Airplanes with tandem seating normally have a limitation requiring solo flight to be made from the front seat in some airplanes or the rear seat in others. Some of the smaller helicopters also require solo flight be made from a specific seat, either the right, left, or center. These seating limitations are noted by a placard, usually on the instrument panel, and they should be strictly followed.

As an aircraft ages, its weight usually increases due to debris and dirt collecting in hard-to-reach locations and moisture absorbed in the cabin insulation. This increase in weight is normally small, but it can be determined only by accurately weighing the aircraft.

Changes of fixed equipment may have a major effect upon the weight of the aircraft. Many aircraft are overloaded by the installation of extra radios or instruments. Fortunately, the replacement of older, heavy electronic equipment with newer, lighter types results in a weight reduction. This weight change, however helpful, can cause the CG to shift, which must be computed and annotated in the weight and balance record.

Repairs and alterations are the major sources of weight changes. It is the responsibility of the FAA-certificate mechanic or repairman making any repair or alteration to know the weight and location of a change, to compute the CG, record the new empty weight and EWCG in the aircraft weight and balance record, and update the equipment lists.

If the newly calculated EWCG should happen to fall outside the EWCG range, it is necessary to perform an adverseloading check. This requires a forward and rearward adverse-loading check and a maximum weight check. These weight and balance extreme conditions represent the maximum forward and rearward CG position for the aircraft. An adverse-loading check is a deliberate attempt to load an aircraft in a manner that creates the most critical balance condition and still remains within the design CG limits of the aircraft. If any of the checks fall outside the loaded CG range, the aircraft must be reconfigured or placarded to prevent the pilot from loading the aircraft improperly. It is sometimes possible to install a fixed ballast in order for the aircraft to operate again within the normal CG range.

The FAA-certificated mechanic or repairman conducting an annual or condition inspection must ensure the weight and balance data in the aircraft records is current and accurate. It is the responsibility of the PIC to use the most current weight and balance data when operating the aircraft.

Stability and Balance Control

Balance control refers to the location of the CG of an aircraft. This is of primary importance to aircraft stability, which is a factor in flight safety. The CG is the point at which the total weight of the aircraft is assumed to be concentrated, and the CG must be located within specific limits for safe flight. Both lateral and longitudinal balance are important, but the prime concern is longitudinal balance; that is, the location of the CG along the longitudinal or lengthwise axis.

An airplane is designed to have stability that allows it to be trimmed to maintain straight-and-level flight with hands off the controls. Longitudinal stability is maintained by ensuring the CG is slightly ahead of the center of lift. This produces a fixed nose-down force independent of the airspeed. This is balanced by a variable nose-up force, which is produced by a downward aerodynamic force on the horizontal tail surfaces that varies directly with the airspeed. If a rising air current should cause the nose to pitch up, the airplane slows and the downward force on the tail decreases. The weight concentrated at the CG pulls the nose back down. If the nose should drop in flight, the airspeed increases and the increased downward tail load brings the nose back up to level flight.

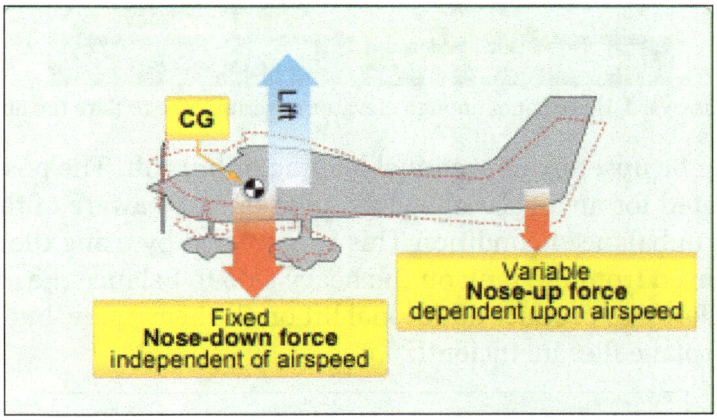

Longitudinal Forces Acting on an Airplane in Flight

As long as the CG is maintained within the allowable limits for its weight, the airplane has adequate longitudinal stability and control. If the CG is too far aft, it is too near the center of lift; the airplane is unstable and difficult to recover from a stall. If the unstable airplane should enter a spin, the spin could become flat making recovery difficult or impossible. If the CG is too far forward, the downward tail load needs to be increased to maintain level flight. This increased tail load has the same effect as carrying additional weight; the aircraft must fly at a higher angle of attack and drag increases.

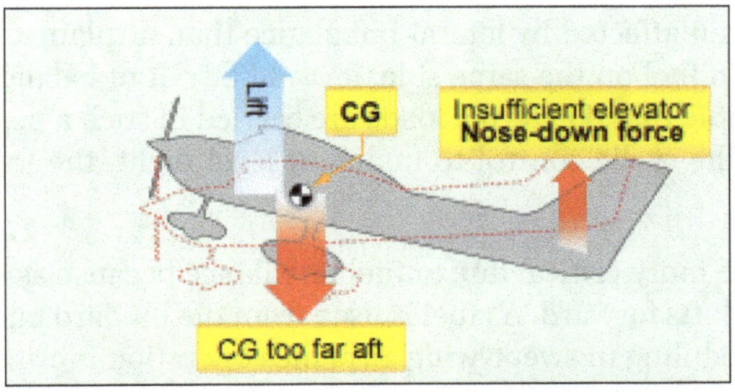

If the CG is too far aft at the low stall airspeed, there might not be enough elevator nose-down authority to get the nose down for recovery.

A more serious problem caused by the CG being too far forward is the lack of sufficient elevator authority. At low takeoff speeds, the elevator might not produce enough nose-up force to rotate; on landing there may not be enough elevator force to flare the airplane. Both takeoff and landing runs are lengthened if the CG is too far forward. The basic aircraft design is such that lateral symmetry is assumed to exist. For each item of weight added to the left of the center line of the aircraft (also known as buttock line zero or BLo), there is generally an equal weight at a corresponding location on the right.

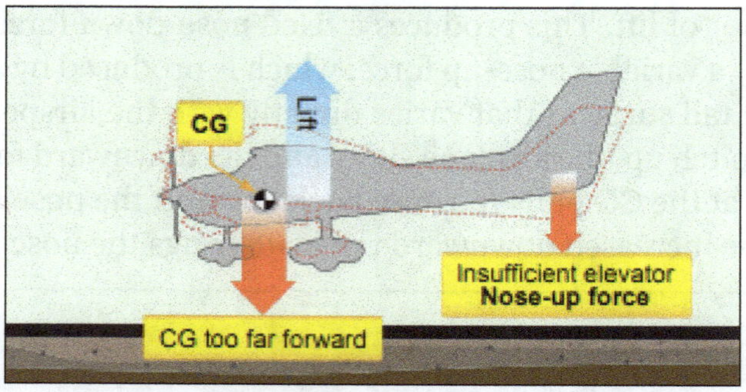

If the CG is too far forward, there is not enough elevator nose-up force to flare the airplane for landing.

The lateral balance can be upset by uneven fuel loading or burnoff. The position of the lateral CG is not normally computed for an airplane, but the pilot must be aware of the adverse effects that result from a laterally unbalanced condition This is corrected by using the aileron trim tab until enough fuel has been used from the tank on the heavy side to balance the airplane. The deflected trim tab deflects the aileron to produce additional lift on the heavy side, but it also produces additional drag, and the airplane flies inefficientl.

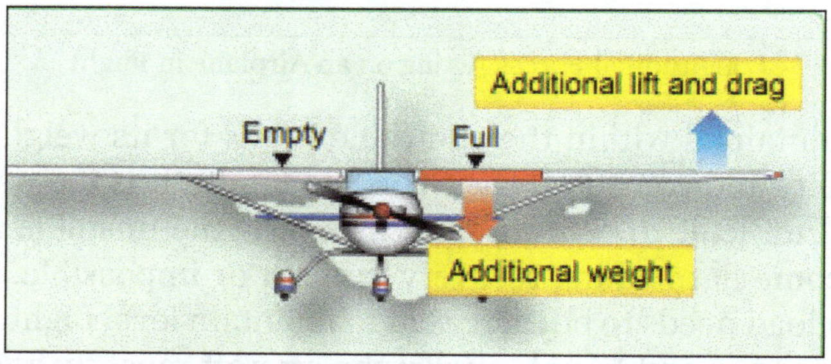

Lateral imbalance causes wing heaviness, which may be corrected by deflecting the aileron. The additional lift causes additional drag, and the airplane flies inefficiently

Helicopters are more often affected by lateral imbalance than airplanes. If a helicopter is loaded with heavy occupants and fuel on the same side, it could be out of balance enough to make it unsafe to fly. It is also possible that if external loads are carried in such a position that requires large lateral displacement of the cyclic control to maintain level flight, the fore-and-aft cyclic control effectiveness is limited.

Swept-wing airplanes are more critical due to fuel imbalance because as the fuel is used from the outboard tanks, the CG shifts forward. As fuel is used from the inboard tanks, the CG shifts aft. For this reason, fuel-use scheduling in swept-wing airplanes operation is critical.

Weight Control for Aircraft other than Fixed and Rotor-wing

Some light aircraft utilize different methods of determining weight and balance from the traditional fixed and rotor-wing aircraft. These aircraft achieve flight control by methods different from the fixed-wing airplane or helicopter. Most notable of these are weight-shift control (WSC) aircraft (also known as trikes), powered parachutes, and balloons. These aircraft typically do not specify either an EWCG or a CG range. They require only a certified or approved maximum weight. To understand why this is so, a look at how flight control is achieved is helpful.

Airplanes and WSC aircraft control flight under the influenc of the same four forces (lift, gravity, thrust, and drag), and around the same three axes (pitch, yaw, and roll). However, each aircraft accomplishes this control in a very different manner. This difference helps explain why the fixed-wing airplane requires an established weight and a known CG, whereas the WSC aircraft only requires the known weight.

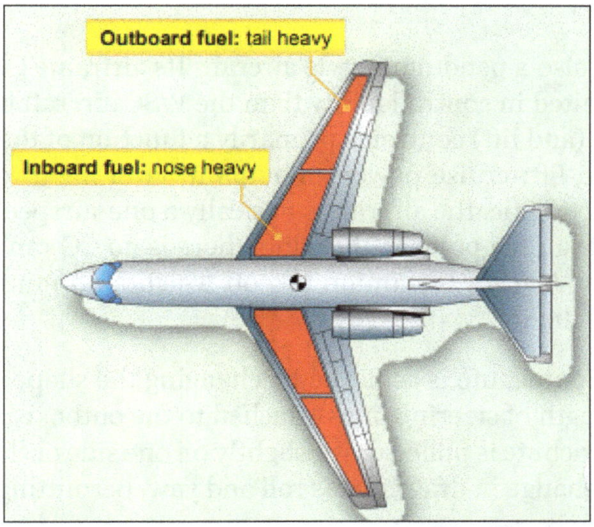

Fuel in the tanks of a swept-wing airplane affects both lateral and longitudinal balance. As fuel is used from an outboard tank, the CG shifts forward.

The fixed-wing airplane has movable controls that alter lift on various airfoil surfaces to vary pitch, roll, and yaw. In turn, these changes in lift affect the characteristics of the flight parameters. Weight normally decreases in flight due to fuel consumption, and the airplane CG changes with this weight reduction. An airplane utilizes its variable fligh controls to compensate and maintain controllability through the various flight modes and as the CG changes. An airplane has a CG range or envelope within which it must remain if the flight controls are to remain effective and the airplane safely operated.

Weight-Shift Control Aircraft

The WSC aircraft has a relatively set platform wing without a tail. The pilot achieves control by shifting weight. In the design of this aircraft, the weight of the airframe and its payload is attached to the wing at a single point in a pendulous arrangement. The pilot, through the flight controls, controls the arm of this pendulum and thereby controls the aircraft. When a change in flight parameter is desired, the pilot displaces the aircraft's weight by the appropriate distance and direction. This change momentarily disrupts the equilibrium between the four forces acting on the aircraft. The wing, due to its inherent stability, then moves appropriately to reestablish the desired relationship between these forces; the wing flexes and alter its shape. As the shape is changed, lift is varied at different points on the wing to achieve the desired flight parameters The flight controls primarily affect the pitch-and-roll axes. Since there is no vertical tail plane, there is minimal or no yaw control. Unlike in an airplane, the CG experienced by the WSC aircraft wing remains constant. Since the weight of the airframe acts through a single point (the wing attach point), the range over which the weight may act is fixed at the pendulum arm or length. Even though weight decreases as fuel is consumed, weight remains focused at the wing attach point. Because the range is fixed, there is no need to establish a calculated range. The pilot should consult the POH or Aircraft Operating Instructions (AOI) for maximum takeoff weight and minimum and maximum seat weight limits prior to each flight.

Powered Parachutes

The powered parachute is also a pendulum-style aircraft. Its airframe CG is fixed at the pendulum attach point. It is more limited in controllability than the WSC aircraft because it lacks an aerodynamic pitch control. Pitch (and lift) control is primarily a function of the power control. Increased power results in increased lift; cruise power amounts to level flight; decreased power causes a descent. Due to this characteristic, the aircraft is basically a one-airspeed aircraft. Once again, because the CG is fixed at the attach point to the wing, there is no CG range. As with WSC, the pilot should consult the POH or AOI for maximum takeoff weight and minimum and maximum seat weight limits prior to each flight.

Roll control on a powered parachute is achieved by changing the shape of the wing. The change is achieved by varying the length of steering lines attached to the outboard trailing edges of the wing. The trailing edge of the parachute is pulled down slightly on one side or the other to create increased drag along that side. This change in drag creates roll and yaw, permitting the aircraft to be steered.

Balloons

The balloon is controlled by the pilot only in the vertical dimension; this is in contrast to all other aircraft. He or she achieves this control through the use of lift and weight. Wind provides all other movement. The CG of the gondola remains constant beneath the balloon envelope. As in WSC and powered-parachute aircraft, there is no CG limitation.

Underestimating the Importance of Weight and Balance

Many pilots, from sport pilot to commercial pilot, tend to underestimate the importance of proper weight and balance of their aircraft. Load sheets are taken for granted and hasty calculations are made of the aircraft's CG. Unfortunately, each year there are a number of accidents related to weight and balance issues. Many of these occurrences could have been avoided had more attention been given to weight and Every student pilot is taught how to work a weight and balance problem and that it is important to make sure every flight is loaded "within the envelope" (no more than maximum gross weight) for both takeoff and landing. But does he or she really understand just why this is so and the disastrous effect of being out of the envelope? Two examples of documented cases are provided below in an effort to indicate the serious nature of maintaining the proper weight and balance. In case studies when weight and balance was listed as the major factor of the accident, many were fatal.

For instance, a small aircraft was loaded with hunters, gear, and dogs (none of the dogs were secured inside the aircraft). During takeoff, all the dogs went to the aft of the airplane. This shifted the CG well aft of its allowable limit. The airplane stalled and crashed. The airplane was destroyed with casualties.

Another accident occurred when a group of skydivers were sitting on the floor toward the aft portion of the airplane (they were unsecured). During takeoff, the CG was again well beyond its aft limit. The airplane stalled and crashed. The airplane was destroyed with casualties.

There is a safety factor built into the formula for maximum gross weight. Any airplane can fly when it takes off at a weight greater than maximum gross weight if the runway is long enough and the density altitude is low enough. However, landing is a different matter. All airplanes are built to withstand an occasional hard landing, but what would happen if the hard landing were combined with a substantially overweight airplane? Something would probably break at that time or the structure would be weakened enough to break sometime in the future when everything might seem normal to a pilot unaware of the previous situation. Even more disastrous than an overweight, hard landing is reaching or exceeding the structural integrity of the metal and/or composite design values when maneuvering or when turbulence is encountered. Hidden damage could result, causing an unexpected catastrophic failure at some future time.

If an airplane is certificated with a maximum gross weight of 6,000 pounds (its weight on the ground) and is rolled into a 60° bank, the forces exerted make it feel as if it weighed 12,000 pounds. At its maximum certificated gross weight, there is no problem because the aircraft is operated within its certificated maneuvering loads. But loaded to 8,000 pounds with a 60° bank or an abrupt pullup, it suddenly weighs 16,000 pounds and might not be able to perform! Even if it could, there would probably be internal stress damage that would show up on future flights.

How to Calculate Aircraft Weight and Balance

Step 1: Find your Aircraft's Empty Weight and Moment

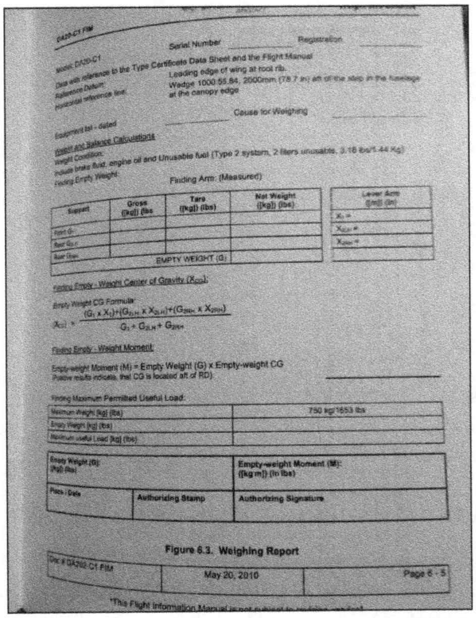

DA20-C1 FIM

Serial Number ______ Registration ______

Model: DA20-C1

Data with reference to the Type Certificate Data Sheet and the Flight Manual

Reference Datum: Leading edge of wing at root rib.

Horizontal reference line: Wedge 1000:55.84, 2000mm (78.7 in) aft of the step in the fuselage at the canopy edge

Equipment list - dated ______ Cause for Weighing ______

Weight and Balance Calculations

Weight Condition:

Include brake fluid, engine oil and Unusable fuel (Type 2 system, 2 liters unusable, 3.18 lbs/1.44 Kg)

Finding Empty Weight: Finding Arm: (Measured)

Support	Gross ([kg]) (lbs	Tare ([kg]) (lbs)	Net Weight ([kg]) (lbs)
Front G_1			
Rear G_{2LH}			
Rear G_{2RH}			
		EMPTY WEIGHT (G)	

Lever Arm
X_1 =
X_{2LH} =
X_{2RH} =

Finding Empty - Weight Center of Gravity (X_{CG}):

Empty Weight CG Formula:

$$(X_{CG}) = \frac{(G_1 \times X_1)+(G_{2LH} \times X_{2LH})+(G_{2RH} \times X_{2RH})}{G_1 + G_{2LH} + G_{2RH}}$$

Finding Empty - Weight Moment:

Empty-weight Moment (M) = Empty Weight (G) x Empty-weight CG

(Positive results indicate, that CG is located aft of RD)

Finding Maximum Permitted Useful Load:

Maximum Weight [kg] (lbs)	750 kg/1653 lbs
Empty Weight [kg] (lbs)	
Maximum useful Load [kg] (lbs)	

Empty Weight (G): [kg] (lbs)		Empty-weight Moment (M): ([kg m]) (in lbs)
Place / Date	Authorizing Stamp	Authorizing Signature

Figure 6.3. Weighing Report

Doc # DA202-C1 FIM	May 20, 2010	Page 6 - 5

Each aircraft's weight and moment are different. This information will be found in the Pilot's

Operating Handbook (POH) specific to the aircraft you are flying. It will be in the weight and balance chapter which is standard chapter six.

Step 2: Find Weights of Everything Loaded

This includes pilot, all passengers, baggage, and the fuel that will be loaded into the aircraft. You will find the fuel load in the first chapter of the POH. Keep in mind that you will use the amount labeled "usable fuel" because unusable fuel is included in the aircraft's weight and that fuel weighs six pounds per gallon.

Step 3: Calculate the Moment for the Load

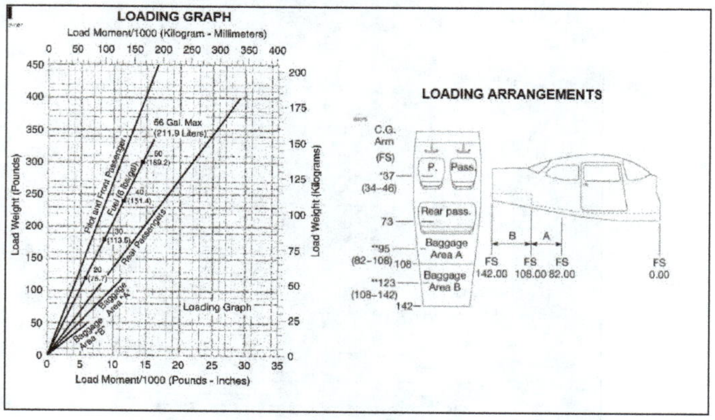

The POH you will find either a chart similar to the one on the left or a figure like the one shown on the right. If you have the chart you can follow the weight to the appropriate line to find the moment. If you have the figure, you need to multiply the weight of each position by the number given as the arm to find the moment. Calculate the moments for the pilot, all passengers, fuel, and baggage.

Example:

Using the graph above, if the pilot and front passenger weighed 300 pounds, you would find 300 on the vertical axis. You then follow that over to where it intersects the line labelled "Pilot and front passenger". You then follow that down to the horizontal axis to get your moment value of about 11. Using the other method with the same 300 pounds you would multiply by the value given for the pilot and front passenger which gives you 300*37=11,100

Step 4: Total Up the Weights and the Moments

Add together the weights for the aircraft, people, fuel, and baggage. Also add the moments found in the previous step (pilot, passengers, fuel, and baggage) to the moment of the empty aircraft.

Step 5: Determine if Center of Gravity is within Limits

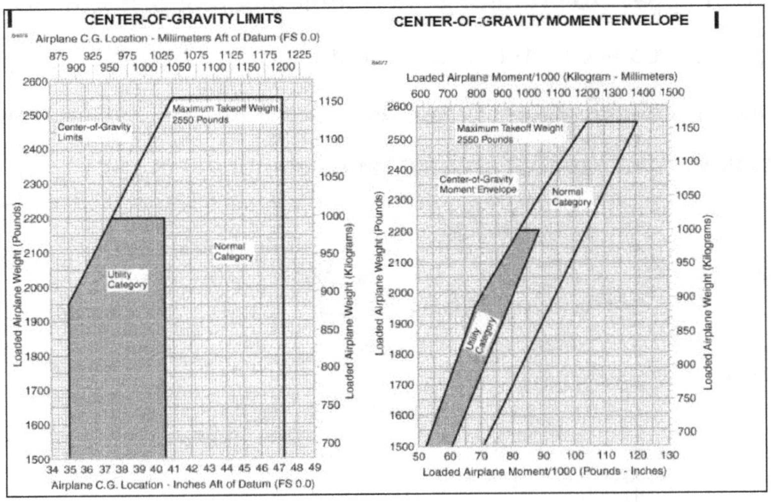

In chapter 6 of the POH find the chart that shows the acceptable limits of the center of gravity. It will look like one of the charts pictured above. One compares the total weight to the total moment and one compares the total weight to the arm of the center of gravity. If your chart uses the arm all you need to do is divide the total moment by the total weight to find that value. If the intersection falls within the area outlined on the chart you are within acceptable limits to fly.

Example

If your total weight was 2500 and total moment was 105 (105,000) you would find these values on their specific axes on the second chart pictured and their intersection lies within the outlined area. Using the arm you would find 105,000/2500=42 which when found on the first chart also falls withing the border.

How to Test and Troubleshoot Aircraft Equipments

Troubleshooting Skills

Three general skills that appear to produce good troubleshooting performance:

- Ability to repair or replace system components.
- Ability to make tests to eliminate components from further consideration.
- Ability to employ a strategy to search for failed components.

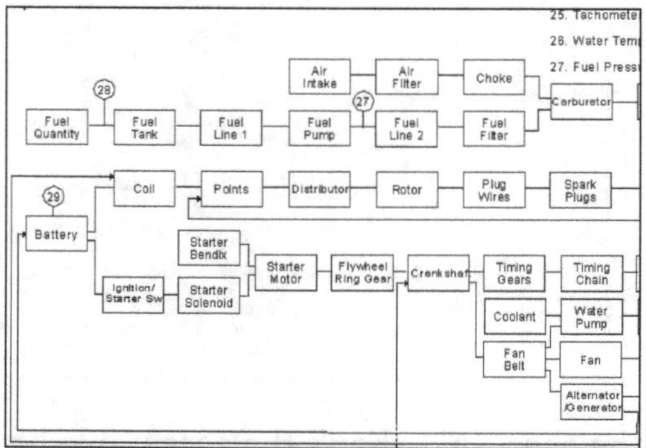

A logical troubleshooting diagram for an automobile engine

After-Maintenance Verification

We normally think of testing as an activity aimed at identifying problems and their causes. From this perspective, we expect testing to reveal a symptom or a cause for an observed symptom. However, there is one type of test in which we expect to find nothing wrong. After we perform maintenance on a component or system, we expect it to be fully functional. To verify that we have, indeed, left the system in a functional state, we normally perform some type of post-maintenance verification test. Such tests are sometimes called quality assurance tests.

An interesting aspect of both preventative maintenance (PM) and after-maintenance testing is that it is quite possible to test a component or system until it breaks. We are never really sure whether or not a particular test has left the system in a functional state. For example, a component could pass an electronic test, but the test itself could degrade the component's operation. Thus, there is a tradeoff between the amount of testing we should perform and the ultimate reliability of the components we are maintaining.

Algorithm

Algorithm is a fancy term for a method or a procedure used to solve a problem. In aviation maintenance, the "problem" is some type of failure in a system or a component. Troubleshooting algorithms are usually based on logic, although they don't have to be.

Automated Test Equipment (ATE)

In the early days of aviation, aircraft were relatively simple and maintenance was straightforward. As aircraft and their components became more sophisticated, diagnosis and repair became more complex. As part of this evolution, automated test capabilities became part of the maintenance technicians "toolbox." Automated test equipment (ATE) describes a broad range of partially- or fully-automated test capabilities.

There is no universally-accepted definition of ATE. The equipment spans the range from extremely sophisticated and expensive test consoles to simple equipment that performs programmed checks on a single avionics module's output. All ATE is either set up to perform a series of tests without requiring human intervention, or can be programmed to do so.

Built-In Test Equipment (BITE)

As with ATE, Built-In Test Equipment (BITE) has evolved with the increased sophistication of aircraft subsystems. BITE is included in the design of a component, module, or subsystem. Except for that common thread, however, BITE describes an extremely broad range of equipment complexity and sophistication.

The simplest BITE might allow a technician or flight crew member to perform a "go/no-go" self-test on a specific module. A more-sophisticated module might allow a technician to perform tests with built-in switches and indicators on a module. The most-sophisticated BITE retains performance information over a number of flight legs, diagnosing failures and displaying specific instructions to technicians.

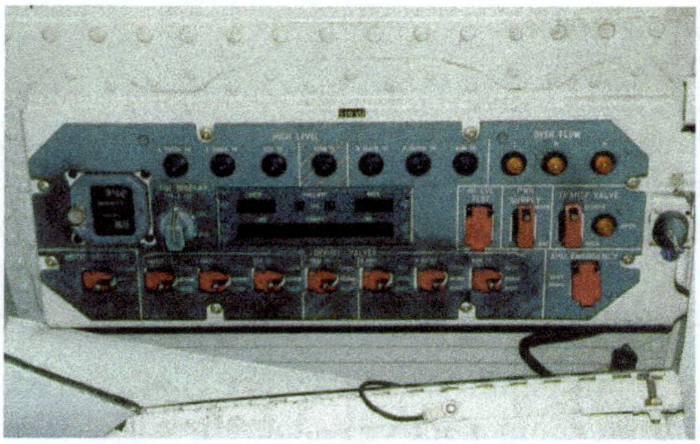

Example of a readout unit for built-in test equipment

Certain BITE is capable of reconfiguring systems to maintain functional capabilities despite component failures. A good example is the control system on the General Motors Northstar engine which can change valve timing and cylinder firing order to compensate for a complete loss of coolant. This BITE capability has been discussed as a way to allow combat aircraft to "self-repair" in battle.

Cannot Duplicate (CND)

Cannot Duplicate (CND) commonly describes a situation in which AMTs cannot cause equipment to exhibit the same failure symptoms other AMTs or flight crew members reported. CNDs are a concern in aviation maintenance because of the high proportion of LRUs classified as "CND" during depot repair. Between 30 and 60 percent of all LRUs pulled from aircraft and sent to depots for repair are eventually tagged as CND.

The inability to reproduce failure symptoms does not necessarily mean that reports of those symptoms were spurious. Complex equipment such as inertial navigation modules often exhibits intermittent failures, i.e., glitches. However, CNDs cause repair processes to be inefficient. If 50 percent of the LRUs returned to a depot are CNDs, approximately half of the technicians' time is spent working on modules that will not be repaired.

Consistent Fault Set (CFS)

Consistent Fault Set is one of the names given to the group of all possible failures that can reasonably explain a given set of trouble symptoms. The name comes from the fact that the group contains faults "consistent" with the symptoms. For example, when an automobile won't start, the CFS could contain an ignition failure, but would not contain a failed seat back adjustment control.

Decision Tree

One type of maintenance job performance aid (JPA) is called a "decision tree". A decision tree is a printed or computerized chart that directs the maintenance technician along a logical testing and diagnosis path for a particular system or product. After each test or observation, the decision tree branches to another test (or conclusion) based on the test results. An easy characterization of a decision tree is a series of "if-then" statements, e.g. "If the voltage is below 'x,' then do this."

Depot Maintenance

LRUs thought to be non-functional are sent to a depot for repair. Depots are shops set up to perform extensive diagnosis on particular types of LRUs. For example, avionics modules are typically sent to a depot set up to handle them. Depot technicians typically have available various types of test and calibration equipment, as well as spare parts for the modules they work on.

Since depot technicians work on the same types of modules each day, they develop significant expertise related to these modules. Their experience and the availability of sophisticated test equipment tends to make depot maintenance very efficient.

Easter Egging

One method of troubleshooting is to replace various LRUs until the symptoms of trouble disappear. This method is known as "Easter Egging" because a technician never really knows where he or she will find the failed part. Easter Egging is an extremely inefficient, expensive way to find a problem.

Einstellung (Psychic Blindness)

Einstellung describes a phenomenon discovered in the early 1940's and since shown to exist in different domains. Researchers have found that when people have spent time solving one particular type of troubleshooting problem, it is virtually impossible for them immediately to diagnose a different type of problem. This phenomenon holds even when people are told that they will see a new and different type of malfunction.

Expert System

Expert systems are diagnostic decision-making aids used in a number of different domains, including medicine, geological exploration, and maintenance. Expert systems are usually computer-based. They are generally developed by embedding a system of rules acquired from human experts. For example, if we were developing an expert system for diagnosing problems in aircraft braking systems, we would first try to determine how human experts do such diagnosis and then put these "rules" into our expert system.

Expert systems in aviation maintenance are commonly embedded in computer-based training systems or diagnostic equipment. Students using such training systems can ask an embedded expert for advice at any point in their troubleshooting practice.

Fidelity

Fidelity describes how closely a reproduction resembles the actual (or original) item. For example, musical recordings are often rated in terms of how much they sound like the original performance. In maintenance, *fidelity* describes how closely a simulator or mockup resembles the real item. There are several types of fidelity, e.g., physical, functional, operational, and psychological. Moving-base cockpit simulators are examples of full-fidelity reproductions because they are essentially indistinguishable from the real thing.

Many studies have shown that full physical fidelity is not necessary for training troubleshooting skills. The operational, or psychological, fidelity of simulators used to teach troubleshooting is more important. It is not necessary to move a physical switch if the simulation depicts that switch and allows the student to alter its position.

Half-Split Test

Theoretically, the most efficient test a maintenance technician can perform is the one that provides the most information. Early troubleshooting research involving electronic circuits found that the "best" test is the one eliminating roughly half the items from a set of possibly failed components. Such a test is called a "half-split" test.

Aiding

Aiding means helping human workers perform their job tasks. In any sufficiently complex system, it is possible for humans to become overwhelmed by their job requirements. Once humans approach their mental or manual limits, performance begins to degrade. Common causes of task overload in the maintenance environment include extreme time pressure, attempting to interpret very complex information, or diagnosing multiple failures in complicated aircraft subsystems.

Aiding is common for complex manual control tasks such as piloting an aircraft. In maintenance, aiding usually takes one of two forms:

1. Expert advice in computer-based training or testing systems.
2. Automated interpretation of complex test results.

Successful implementation of aiding requires a thorough understanding of the information requirements of each task. Specific design of aiding must be preceded by some form of task or job analysis.

Failure Modes and Effects Analysis

Failure modes and effects analysis (FMEA) is typically used as a safety analysis technique, using a diagrammatic representation of a particular system as its base. Each system component is presumed to fail in one of a number of modes. The effects of that failure are then determined by analyzing the connections between the failed component and other parts of the system. FMEA is a "bottom up" analysis technique; first we choose a component and then we determine what happens when it fails. Fault tree analysis is the opposite of FMEA.

For testing and troubleshooting, FMEA is most useful for generating and testing hypotheses concerning failure symptoms. For example, if we think a failed avionics module might be causing certain symptoms, we can use FMEA to determine what effects a failure of that module would have on the overall system. In other words, we want to know if the symptoms are consistent with a specific failure mode for a particular component?

Fault Tree Analysis

Fault tree analysis (FTA) is opposite of FMEA. It shares with FMEA its use of a diagrammatic representation of a system, but FTA is much more an exercise in logic. For FTA, we first specify a particular set of symptoms then logically determine every component failure that could cause those symptoms. FTA-like methods in aviation (and other) maintenance are typically used as troubleshooting guides. A troubleshooting guide generally lists a series of symptoms, e.g., landing gear green light will not illuminate, then directs the troubleshooter to a number of possible causes for the symptoms.

FTA's biggest shortcoming is that it must include *every* possible cause of the observed symptoms. Logically, being all-inclusive is quite difficult. However, when FTA is used in conjunction with FMEA, it quickly narrows the search for a consistent fault.

Probabilistic Risk Assessment (PRA)

The essence of troubleshooting is identifying the failure(s) causing a particular set of symptoms. In nearly all physical systems, some failures are more likely than others. Manufacturers often test their products to determine the failure rates of particular components. When component failure rates are known, or can be reasonably estimated, they can also be mathematically combined to predict the probability of various types of system failures. This process is known as Probabilistic Risk Assessment (PRA).

PRA is commonly used to assess safety risks associated with complex systems, such as nuclear power plants, or of potentially dangerous operations, such as extended twin-engine operations over water (ETOPS). In the maintenance domain, PRA can determine the most likely failure modes for a particular system, and the information can then be factored into troubleshooting guides or expert systems.

Proceduralization

Proceduralization can improve troubleshooting performance. When troubleshooting is properly proceduralized, performance differences between expert and novice troubleshooters can be virtually eliminated. As with any other endeavor, however, there are good and bad procedures. More accurately, there are procedures improving performance and procedures with either little effect or that actually degrade performance.

Proceduralization must be preceded by a thorough analysis of relevant troubleshooting tasks to determine what each troubleshooting step tries to accomplish, what information is required and produced, and what tests or tools should be used. As with certain aspects of automation, it is possible to proceduralize to the extent that human technicians are left with an essentially mechanical role. We should also note that troubleshooting procedures exist in an overall organizational and work environment. Good procedures are worthless if they are used improperly or ignored.

Simulation-oriented Computer-based Instruction

Teaching testing and troubleshooting skills has traditionally been problematic for maintenance organizations. Traditional training methods such as classroom lectures tend to be expensive and inefficient. Simulation-Oriented Computer-Based Instruction (SOCBI) has a number of advantages for the effort to teach troubleshooting skills. SOCBI capitalizes on two findings:

1. Troubleshooting training most efficiently addresses specific tasks related to the actual equipment that students encounter on the job.

2. Developing troubleshooting skills requires a great deal of practice, preferably on a representation of actual equipment.

SOCBI meets both requirements by exposing students to realistic failures on a computer simulation of the actual systems they will see on the job. SOCBI's focus is on high psychological and visual fidelity: students perform tests and acquire information much as they would with real equipment. In a number of studies, SOCBI has been shown to produce troubleshooting performance equal to, or better than, more traditional, less efficient methods.

Training Approaches that Improve Troubleshooting Performance

- Using simulators and mockups
- Simulation-Oriented Computer-Based Instruction
- Almost any training method that allows meaningful practice
- Providing heuristics (rules-of-thumb)
- Teaching context-specific troubleshooting procedures
- Teaching how to analyze failure symptoms
- Teaching troubleshooting algorithms, such as half-splitting, when combined with training in
- how to generate possible failure hypotheses

While it is difficult to teach troubleshooting skills, it is certainly possible. There are a number of worthwhile approaches to troubleshooting; many of these were developed in the aviation maintenance environment.

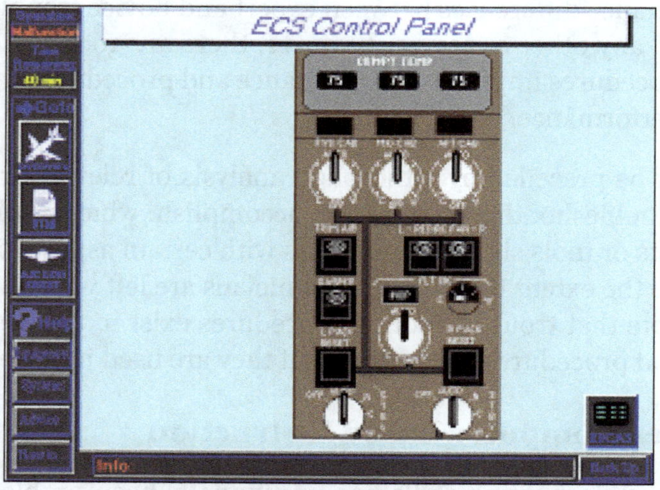

Figure: SOCBI depiction of an ECS control panel

Simulation-Oriented Computer-Based Instruction. Simulation-Oriented Computer-Based Instruction (SOCBI) is one of the most diligently-studied training methods that combines many elements for success in troubleshooting training. Work in SOCBI began in the aviation maintenance domain in the late 1970's. SOCBI provides students with a two-dimensional, interactive depiction of the particular system or component that they are learning to troubleshoot. If the component is small enough, an SOCBI module can actually show a picture of its controls and displays, for example, the environmental control system (ECS) panel shown in figure.

Students use the working controls and displays to practice diagnosing a number of (usually randomly occurring) faults built into the simulation. SOCBI modules also contain diagrammatic, i.e., logical, representations of the system being taught. An example of these diagrams is shown in figure. These functional/logical diagrams teach students how a system is functionally connected and allow them to use logical troubleshooting algorithms such as half-splits.

Effective SOCBI allows students to acquire diagnostic information from the same sources available in the work environment. Students must be able to observe indications, such as lights and gauges; to perform specific tests on the system; to receive verbal reports from flight crew members, etc.

A number of SOCBI systems have been compared with more traditional training methods such as classroom instruction and demonstrations of actual equipment. In these studies, SOCBI produces troubleshooting performance as good as, or better than, that produced by less-efficient techniques.

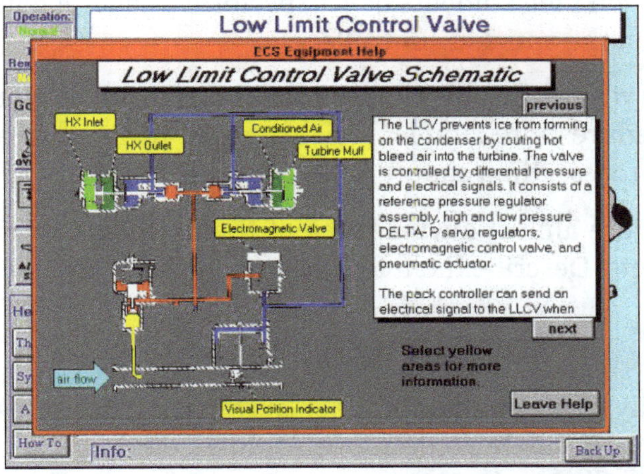

Figure: Example of a functional/logic diagram from an SOCBI module

Practice. Regardless of which training method, or combination of methods, one uses to teach troubleshooting skills, students must be given an opportunity for practice. In table, we have explicitly specified that troubleshooting practice must be meaningful and listed the most important components of meaningful practice.

Components of Meaningful Troubleshooting Practice

To be meaningful, troubleshooting practice should:

- Pertain to the equipment that will actually be maintained on the job
- Be done using mockups that provide the same types of information as the real system
- Allow students to gather information from the same sources as in the actual work environment

- Provide feedback regarding the outcome of various tests and other actions
- Allow students to know how long their actions would take in the actual work environment

Context-specific knowledge. Many maintenance skills are generalizable from one domain to another. For example, skill in the use of tools for repairing automobile engines is directly applicable to using tools to repair turbine engines. However, troubleshooting skills tend to be context-specific. The ability to diagnose problems with a television set does not directly transfer to troubleshooting avionics modules. When teaching troubleshooting knowledge, it is important to provide specific information. Table lists some elements of context-specific troubleshooting knowledge.

Elements of Context-specific Troubleshooting Knowledge

Troubleshooting knowledge should be:

- Explicit - Tell students how you expect them to use the information you are providing. Don't rely on them to guess how it should be used.
- Specific - Relate troubleshooting steps to the component(s) on which students will be working. For example, don't tell students how to perform a general half-split test. Tell them how to do a half-split on the antiskid controller.
- Simple - Students will not be able to remember long, involved troubleshooting procedures. Break these procedures into simple, serial steps. If there is no easy way to decompose a troubleshooting process, then supply a written procedure.
- Heuristic - There are almost always rules-of-thumb for troubleshooting specific components or subsystems. Describe them for the students.

Evaluating Automation Alternatives

Most issues that generally apply to maintenance automation are relevant to testing and troubleshooting. As aircraft systems become more sophisticated, we can expect more emphasis on automated testing and diagnosis. Newer systems and even individual components depend on software for their functional capabilities. This general trend affects many maintenance domains.

One non-aviation example of the emphasis on software is found in network communication components. Until recently, hubs, routers, switchers, and other network components were essentially hardware-based. Since newer components rely instead on built-in software, the old methods of troubleshooting and repairing these components are inappropriate. While aircraft manufacturers and system vendors have until recently resisted the general movement to software, the trend is clear.

Many modern aircraft components contain sophisticated built-in automated test and diagnostic capabilities. Line maintenance for these modules often consists of reading diagnostic information from a cockpit control-display unit (CDU). One example of such an automated module is the inertial navigation unit on the DC-10 aircraft.

Probably the biggest issue related to maintenance automation, including testing and troubleshooting, is that it is possible to become too dependent on automated equipment. Studies of cockpit automation have consistently shown that, rather than eliminating operational problems, automation is the source of an entire new set of errors.

Reducing Troubleshooting Errors

The goal of every maintenance organization is 100% error-free performance. While this is a worthwhile goal, expecting 100% *anything* is unrealistic. In fact, one of the few rules on which human factors professionals agree is that people make mistakes. Most testing and troubleshooting in the aviation industry includes a human element, so we can expect errors to occur.A more realistic goal is to *reduce* troubleshooting errors as much as possible. An adjunct goal should be to make errors, when they occur, as obvious and benign as feasible.

Past research and experience have shown a number of ways to reduce testing and troubleshooting errors. Common error-reduction methods include the following:

- Automate
- Instruct in algorithms or heuristics
- Proceduralize
- Practice
- Reduce Complexity
- Relax time pressure
- Select technicians
- Work in teams.

Guidelines for Selecting Testing and Troubleshooting Automation

Automation for testing and troubleshooting should include the following characteristics:

- Very few "modes" - The number of different operating modes should be kept to a minimum, i.e., no more than 3 or 4.
- Clearly displayed mode - The current operating mode of the automated equipment should be clearly displayed to technicians.
- Explicit actions - The automated equipment should inform technicians regarding what it is doing (and why, if the reason is not obvious).
- Common terminology - The terminology used in the technician-equipment interface should be compatible with the terminology used for non-automated troubleshooting tasks.
- Consistent interface - The technician/equipment interface should be consistent for all automated troubleshooting equipment.

- No computer babble - Understanding the information, instructions, labels, etc., should require no software expertise on the part of the operating technician.
- No bit tweaking - Technicians should not have to examine individual bits or bytes of data to acquire troubleshooting information. The user interface should display information in an easily interpretable format.
- Operator override - Allow users to override automated testing or troubleshooting functions at any time.

Steps in Automating Testing and Troubleshooting Tasks

Step activity:

1. Analyze errors
2. Analyze high-error tasks
3. Determine automation candidates
4. Assign manual/automated steps
5. Identify automation alternatives
6. Select automation alternatives
7. Integrate manual and automated steps

Automation does not eliminate all errors during testing and troubleshooting.

Improper automation can actually *introduce* errors that do not typically occur in manual troubleshooting. However, when automation is used properly, it can reduce certain types of maintenance errors. The trick is in knowing where, how, and how much to automate.

The specific steps to take when introducing automation for testing and troubleshooting tasks are summarized in table. They are similar to the steps for introducing any type of automation. The most important step is to analyze the errors and tasks for which automation is anticipated. For automation or any other error-reduction method to work, we must first know relevant information like the following:

- What types of errors are occurring
- What is the frequency of each error type
- In which tasks do errors occur
- Who is committing each type of error
- What are the working conditions in which errors occur
- Is there a connection among errors and time of day, time of year, etc.

After completing the error and task analysis, emphasize properly allocating steps either to human technicians or to automated equipment. Improper allocation does not reduce (and might increase) the number or type of errors. For example, suppose we identify the following two errors:

1. Replacing test leads on the incorrect connector pins.
2. Incorrectly interpreting the display on an eddy current tester.

In this example, automation, perhaps in the form of an expert system should help with the second problem. For the first problem, automation probably would not help. This task requires complex cognitive ability coupled with manual dexterity.

Instruct in Algorithms and Heuristics

Teaching certain theoretical information, such as a theory of operation, doesn't necessarily improve troubleshooting or testing performance. However, teaching heuristics (rules-of-thumb) and particular troubleshooting algorithms can improve the efficiency of testing and troubleshooting, including reducing errors.

Rules-of-thumb vary, depending on the specific component or system. An example of a rule-of-thumb is: "If the symptoms include a low pressure indication, then always check the pressure sender unit first." Rules-of-thumb represent the distilled wisdom of expert technicians who have diagnosed problems over a long period. Embedded expert systems depend on a rule base developed by consulting expert troubleshooters.

Algorithms are usually unwritten procedures telling troubleshooters generally how to proceed. Some research studies show that troubleshooting performance improves when technicians are reminded, in general terms, what they should do first, second, etc. For example, a general algorithm might require a technician to gather information related to failure symptoms, to generate as many hypotheses consistent with the symptoms as possible, to prioritize the hypothesis set, etc. Such general algorithms seem to have the effect of dissuading technicians from deciding on a specific failure being the cause of the symptom before they have enough information.

Proceduralize

Troubleshooting and testing errors can be reduced by proceduralizing a task. Previous studies have shown that procedures are more effective for reducing errors in complex than in simple systems. The notion of using maintenance procedures is certainly not new. Within the aviation maintenance industry, various types of procedures are commonly used. In fact, workcards constitute the most common type of procedure.

A number of issues related to procedures are beyond the scope of this *Guide*. We do not give detailed guidelines related to procedural format, typography, placement of warnings, and other aspects.

Maintenance procedures serve various purposes. Reducing errors is certainly an implicit goal of all such procedures. However, error reduction may not be the only, or even the main, focus of procedures. Other valid reasons to develop procedures include to comply with FAA requirements, to reduce the required experience level of technicians, to reduce the performance variability inherent in certain tasks, etc. Procedures aimed at reducing errors should be developed, stored, and used according to the guidelines listed in Table.

Guidelines for Procedures Aimed at Reducing Testing and Troubleshooting Errors

Procedures aimed at reducing errors should be:

- Specific - Procedures should be written for a specific component, system, or piece of test equipment.
- Clear - The terminology should be consistent with the language commonly used by the people who will complete the procedure.
- Explicit - Tell users what they are supposed to do. Do not depend on technicians to read between the lines.
- Detailed - Include all required steps in the procedure. Don't assume that technicians will know all of the substeps required to achieve a specific system state.
- Accessible - Procedures must be stored in a place and manner so they are easy to obtain.
- Usable - Procedures must exist in a format and on media that make them easy to use while technicians perform the tasks they describe.

Practice

That "practice makes perfect" has been proven for troubleshooting tasks. The major factor distinguishing expert troubleshooters from novices is experience, i.e., practice. Troubleshooting is a complex skill with cognitive (mental) and manual elements. As is true of all such skills, troubleshooting proficiency cannot be attained simply by reading books or by listening to someone explain what to do. Providing opportunities for meaningful practice is a valid, relatively inexpensive method to reduce troubleshooting errors. Table lists guidelines for meaningful practice.

In addition to the elements listed in table, we include two other practice-related issues – fidelity and availability. Troubleshooting practice does not have to be on real equipment; in fact, real equipment is often an inefficient practice medium with the following drawbacks:

- It is difficult to know the precise nature of failures embedded in real equipment
- Experts often disagree as to the appropriate troubleshooting path(s) for failures in real equipment
- Using real equipment as practice aids prevents the equipment from being used to support operations
- Errors made while troubleshooting real equipment can have safety implications
- For failures to be intentionally embedded in real equipment, someone has to embed the failures, check the equipment when practice troubleshooting is complete, and ensure that only controlled failures are present.

The other issue related to practice regards timing and accessibility. Although failures are relatively rare, across an entire fleet of aircraft various failures occur each day. A line or depot technician sees a number of failures over a week or month. Troubleshooting errors tend to occur when technicians see unusual, infrequent problems. Effective practice is conducted on a recurring basis over a long time and is available when a technician's schedule allows.

Use simulation. All these considerations lead to our recommendation that practice aimed at reducing errors be conducted using some type of *computer-based simulation*. Such simulations now exist on laptop computers. Table lists some major advantages of computer-based simulation for error-reduction practice.

Reduce Complexity

Testing and troubleshooting errors tend to increase with the complexity of the system or component being maintained. One simple approach to reducing such errors is to reduce the complexity of the systems being maintained. This approach seems ludicrous on its face, i.e., how does one reduce the complexity of a system that already exists? Also, most maintenance organizations don't actually *design* the systems they maintain. A system's complexity is normally determined during its design.

Advantages of using Computer-based simulation for Error-reduction Practice

- It can be placed in convenient locations so it is accessible to technicians in or near their workplace.
- It can have a game-like quality that appeals to younger technicians (who are likely to be the
- most in need of practice).
- It can present complex failure scenarios in a non-threatening environment. That is, there is no actual cost for errors.
- There are no safety implications associated with troubleshooting errors.
- It provides distributed practice over long periods of time.
- It provides immediate feedback regarding errors, costs, efficiency, etc.
- The user interface can be made consistent among all simulated systems and between this simulation and that used for training.

There are two approaches to reducing complexity that can improve testing and troubleshooting performance, i.e., reduce errors:

1. Simplify the testing and troubleshooting *equipment* and *procedures*.
2. Break complex testing and troubleshooting tasks into simpler pieces.

Both approaches are probably being used to some extent within most maintenance organizations.

Simplify equipment and procedures. There is no direct link between aircraft components' and systems' complexity and the complexity of related testing and troubleshooting equipment and procedures. Much line maintenance consists of simply reading and following instructions on a control-display unit (CDU). Testing and troubleshooting a complex, sophisticated avionics module might be nothing more difficult than following the instruction, "Remove unit. Replace with new unit."

Maintenance supervisors usually have direct control over the selection and development of equipment and procedures. As the complexity of these elements decreases, testing and troubleshooting errors also decrease.

Decompose complex tasks. Various studies have shown that complex tasks and skills are learned more efficiently when broken down into simpler components. This reasoning can be carried into actual work tasks so that procedures decompose complex tasks into a series of simple individual steps. The effect of such decomposition on performance depends on the resultant steps' complexity.

The drawback of task decomposition is that resultant tasks might be so detailed that technicians loose track of what they are trying to accomplish. For example, we can accomplish the same purpose by telling a technician either to "remove an access cover" or to "remove screw 1, then remove screw 2." The key to success, of course, is to simplify a complex task without losing context and intent.

Relax Time Pressure

Time pressure degrades troubleshooting and testing performance, and nearly all aviation maintenance tasks are subjected to some form of time pressure. Line maintenance technicians are consistently exposed to time pressure because of their need to have aircraft ready for operation. Depot technicians' time constraints are usually measured in days or weeks, instead of line technicians' hours or minutes. Technicians and mechanics working on major structural checks and aircraft re-configurations are also expected to meet time constraints, but their time pressure is similar to that of construction crews, i.e., such projects have milestones and a due date.

There is a limit for relaxing time pressure on troubleshooting tasks in a line maintenance organization. Many line tasks involve unplanned work. When a flight crew reports that a certain component appears to be malfunctioning, technicians must work within the aircraft's planned ground time to diagnose and repair the problem. Maintenance organizations often deal with time constraints by practicing replacement maintenance for unplanned work. They make no attempt at in-depth diagnosis or repair; instead, they identify LRUs that might contribute to the problem and replace them as quickly as possible.

Time Pressure Reduction Procedure for Line Technicians

Step 1 - Perform an analysis to determine the most likely causes for certain functional symptoms. This analysis can be composed of different elements, including, PRA, FMEA, Fault Tree, etc.

Step 2 - Build a troubleshooting decision tree based upon this analysis. The decision tree should specify which tests to perform and which modules to remove or repair given particular time constraints. For example, if 15 minutes until departure, remove and replace modules A, B, and C. If 30 minutes until departure, test module A, if bad, then do this. If good, then do that.

Step 3 - Develop a tagging procedure for modules removed in Step 2. It is possible that modules will be removed without testing them. Others may have been only partially tested.

Step 4 - Develop a depot screening process for incoming modules. Depot technicians should be able to look at the tags and tell what level of testing was done on the flight line. This will allow them to perform an initial screening to determine whether modules actually exhibit any failure symptoms.

This practice of rapid diagnosis and replacement undoubtedly contributes to most organizations' high rate of CNDs. Such false positive findings should be expected. In fact, as more time pressure is brought to bear on line organizations, depot technicians likely find a higher proportion of CNDs. This is not necessarily bad, but the tradeoff is more depot technician time or unplanned operational delays.

In reality, there appears to be only one viable way to reduce time pressure on line technicians. This procedure is outlined in table.

Select Technicians

Certain individual personality traits, such as cognitive style, have been correlated with testing and troubleshooting performance. Theoretically, if we select technicians based on these traits, we improve testing and troubleshooting performance. However, it is difficult to know whether traits such as introversion are practically significant for aviation-related testing and troubleshooting tasks. Much stronger links appear among performance and training, experience, proceduralization, etc.

Even if one were to select technicians based on individual traits, it's not clear that the traits can be measured with enough reliability for them to be valid predictors of performance. What is clear is that some people are better troubleshooters than others. A viable selection criterion that *could* reduce testing and troubleshooting errors might be to select people who have demonstrated exceptional troubleshooting performance on the job. This strategy would establish an elite class of technicians recognized for their troubleshooting ability.

However, using personnel selection as a method for reducing errors is fraught with problems, including its depressing effect on employee morale and its repercussions on union work agreements.

By forming an elite group of troubleshooters, what message are we giving people not selected? By selecting technicians solely on troubleshooting performance, regardless of other skills and experience, how are we affecting seniority, pay grades, etc.? Based on these considerations, we do not recommend personnel selection as an error-reduction technique, although other selection criteria can be used to enhance maintenance productivity. A natural selection process occurs in the workplace so that technicians tend to consult the best troubleshooters on particular systems or modules -- especially for complex failures where troubleshooting errors are more likely.

Work in Teams

To quote an old expression, "Two heads are better than one." Intuition suggests that two or three technicians working together commit fewer testing and troubleshooting errors than a single person working alone. Even if individual team members commit the same number of errors, the chances of detecting and correcting those errors seem greater for a team than an individual.

There certainly are ample precedents for using teams of technicians. In manufacturing, the use of workgroups, quality circles, and other worker teams has proven to be successful in increasing product quality, i.e., reducing manufacturing errors. In maintenance, the use of multiple technicians has had mixed results.

A classic example of having more than one technician perform a task is the *double-checkoff* procedure. Double checkoff procedures require one individual to perform a task and initial each completed step. A second person takes the procedure, ensures that each step has been completed and initials it a second time. NASA, the commercial nuclear industry, and the military use such procedures for safety-critical tasks.

Experience with such team concepts has proved them to be fallible. For double-checkoff procedures, workers tend to become complacent, i.e., the second person often assumes that the first did everything properly. The second checkoff becomes nothing more than an exercise in writing initials. Teams that work to reduce errors have the following characteristics:

1. Each member maintains a perspective that allows him or her to question other team members' work.
2. Each member must have the temperament to accept other members' observations without becoming defensive.

In the aviation maintenance environment, including technicians and inspectors on the same team might work to reduce errors, depending on individual personalities. Workers' acceptance of a team as a viable working unit seems largely to depend on their perception of how much their management demonstrates its support. One way to ensure failure of the team concept is to treat the team as a mechanism for catching inept workers. Once employees have the idea that the purpose of the team is punitive, any performance-enhancing effects are lost.

Chapter 4 Powerplant Mechanic

An aircraft powerplant is a power source such as an engine that helps airplanes and helicopters propel off the ground. The professionals entrusted with the maintenance of these engines are known as powerplant mechanics. The chapter elucidates the ways to become an aircraft engine specialist and aircraft powerplant technician.

Aircraft Power Plant

The airplane engine and propeller, often referred to as the aircraft powerplant, work in combination to produce thrust. The powerplant propels the airplane and drives the various systems that support the operation of an airplane.

Reciprocating Engines

Most small airplanes are designed with reciprocating engines. The name is derived from the back-and-forth, or reciprocating, movement of the pistons. It is this motion that produces the mechanical energy needed to accomplish work. Two common means of classifying reciprocating engines are:

1. by cylinder arrangement with respect to the crankshaft—radial, in-line, v-type or opposed, or
2. by the method of cooling—liquid or air-cooled.

Radial engines were widely used during World War II, and many are still in service today. With these engines, a row or rows of cylinders are arranged in a circular pattern around the crankcase. The main advantage of a radial engine is the favorable power-to-weight ratio.

In-line engines have a comparatively small frontal area, but their power-to-weight ratios are relatively low. In addition, the rearmost cylinders of an air-cooled, in-line engine receive very little cooling air, so these engines are normally limited to four or six cylinders.

V-type engines provide more horsepower than in-line engines and still retain a small frontal area. Further improvements in engine design led to the development of the horizontally-opposed engine.

Opposed-type engines are the most popular reciprocating engines used on small airplanes. These engines always have an even number of cylinders, since a cylinder on one side of the crankcase “opposes” a cylinder on the other side. The majority of these engines are air cooled and usually are mounted in a horizontal position when installed on fixed-wing airplanes. Opposed-type engines have high power-to-weight ratios because they have a comparatively small, lightweight crankcase. In addition, the compact cylinder arrangement reduces the engine's frontal area and allows a streamlined installation that minimizes aerodynamic drag.

The main parts of a reciprocating engine include the cylinders, crankcase, and accessory housing. The intake/exhaust valves, spark plugs, and pistons are located in the cylinders. The crankshaft and connecting rods are located in the crankcase. The magnetos are normally located on the engine accessory housing.

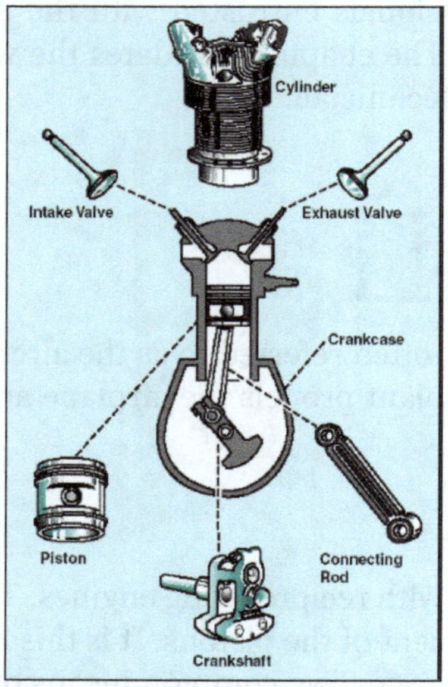

Main components of a reciprocating engine.

The basic principle for reciprocating engines involves the conversion of chemical energy, in the form of fuel, into mechanical energy. This occurs within the cylinders of the engine through a process known as the four-stroke operating cycle. These strokes are called intake, compression, power, and exhaust.

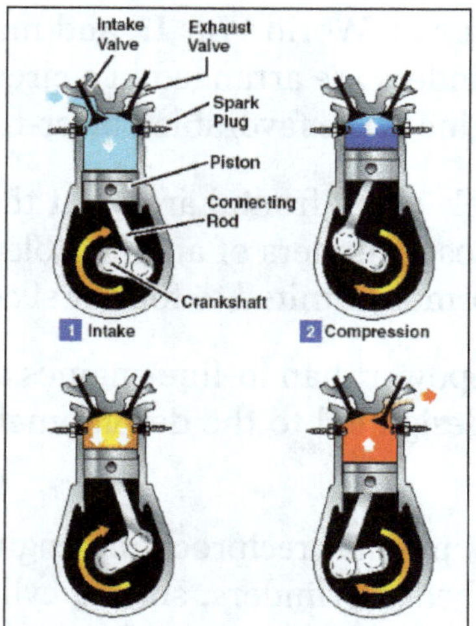

The arrows in this illustration indicate the direction of motion of the crankshaft and piston during the four-stroke cycle.

1. The intake stroke begins as the piston starts its downward travel. When this happens, the intake valve opens and the fuel/air mixture is drawn into the cylinder.
2. The compression stroke begins when the intake valve closes and the piston starts moving back to the top of the cylinder. This phase of the cycle is used to obtain a much greater power output from the fuel/air mixture once it is ignited.
3. The power stroke begins when the fuel/air mixture is ignited. This causes a tremendous pressure increase in the cylinder, and forces the piston downward away from the cylinder head, creating the power that turns the crankshaft.
4. The exhaust stroke is used to purge the cylinder of burned gases. It begins when the exhaust valve opens and the piston starts to move toward the cylinder head once again.

Even when the engine is operated at a fairly low speed, the four-stroke cycle takes place several hundred times each minute. In a four-cylinder engine, each cylinder operates on a different stroke. Continuous rotation of a crankshaft is maintained by the precise timing of the power strokes in each cylinder. Continuous operation of the engine depends on the simultaneous function of auxiliary systems, including the induction, ignition, fuel, oil, cooling, and exhaust systems.

Propeller

The propeller is a rotating airfoil, subject to induced drag, stalls, and other aerodynamic principles that apply to any airfoil. It provides the necessary thrust to pull, or in some cases push, the airplane through the air.

The engine power is used to rotate the propeller, which in turn generates thrust very similar to the manner in which a wing produces lift. The amount of thrust produced depends on the shape of the airfoil, the angle of attack of the propeller blade, and the r.p.m. of the engine. The propeller itself is twisted so the blade angle changes from hub to tip. The greatest angle of incidence, or the highest pitch, is at the hub while the smallest pitch is at the tip.

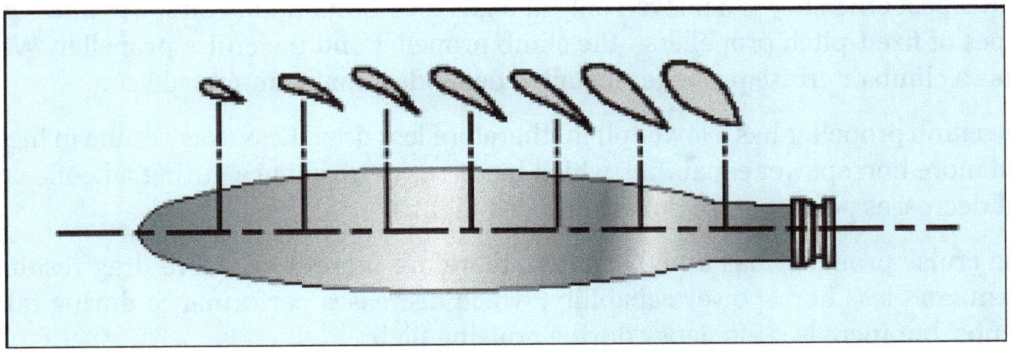

Changes in propeller blade angle from hub to tip.

The reason for the twist is to produce uniform lift from the hub to the tip. As the blade rotates, there is a difference in the actual speed of the various portions of the blade. The tip of the blade travels faster than that part near the hub, because the tip travels a greater distance than the hub in the same length of time.

Changing the angle of incidence (pitch) from the hub to the tip to correspond with the speed produces uniform lift throughout the length of the blade. If the propeller blade was designed with the same angle of incidence throughout its entire length, it would be inefficient, because as airspeed increases in flight, the portion near the hub would have a negative angle of attack while the blade tip would be stalled.

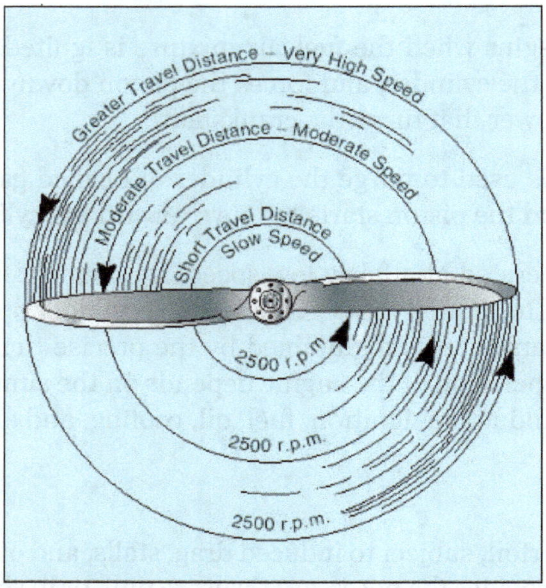

Relationship of travel distance and speed of various portions of propeller blade.

Small airplanes are equipped with either one of two types of propellers. One is the fixed-pitch, and the other is the controllable-pitch.

Fixed-pitch Propeller

The pitch of this propeller is set by the manufacturer, and cannot be changed. With this type of propeller, the best efficiency is achieved only at a given combination of airspeed and r.p.m. There are two types of fixed-pitch propellers—the climb propeller and the cruise propeller. Whether the airplane has a climb or cruise propeller installed depends upon its intended use:

- The climb propeller has a lower pitch, therefore less drag. Less drag results in higher r.p.m. and more horsepower capability, which increases performance during takeoffs and climbs, but decreases performance during cruising flight.
- The cruise propeller has a higher pitch, therefore more drag. More drag results in lower r.p.m. and less horsepower capability, which decreases performance during takeoffs and climbs, but increases efficiency during cruising flight.

The propeller is usually mounted on a shaft, which may be an extension of the engine crankshaft. In this case, the r.p.m. of the propeller would be the same as the crankshaft r.p.m. On some engines, the propeller is mounted on a shaft geared to the engine crankshaft. In this type, the r.p.m. of the propeller is different than that of the engine. In a fixed-pitch propeller, the tachometer is the indicator of engine power.

Engine r.p.m. is indicated on the tachometer.

A tachometer is calibrated in hundreds of r.p.m., and gives a direct indication of the engine and propeller r.p.m. The instrument is color-coded, with a green arc denoting the maximum continuous operating r.p.m.

Some tachometers have additional markings to reflect engine and/or propeller limitations. Therefore, the manufacturer´s recommendations should be used as a reference to clarify any misunderstanding of tachometer markings.

The revolutions per minute are regulated by the throttle, which controls the fuel/air flow to the engine.

At a given altitude, the higher the tachometer reading, the higher the power output of the engine.

When operating altitude increases, the tachometer may not show correct power output of the engine. For example, 2,300 r.p.m. at 5,000 feet produce less horsepower than 2,300 r.p.m. at sea level. The reason for this is that power output depends on air density. Air density decreases as altitude increases. Therefore, a decrease in air density (higher density altitude) decreases the power output of the engine. As altitude changes, the position of the throttle must be changed to maintain the same r.p.m. As altitude is increased, the throttle must be opened further to indicate the same r.p.m. as at a lower altitude.

Adjustable-pitch Propeller

Although some older adjustable-pitch propellers could only be adjusted on the ground, most modern adjustable-pitch propellers are designed so that you can change the propeller pitch in flight. The first adjustable-pitch propeller systems provided only two pitch settings - a low-pitch setting and a high-pitch setting. Today, however, nearly all adjustable-pitch propeller systems are capable of a range of pitch settings.

A constant-speed propeller is the most common type of adjustable-pitch propeller. The main advantage of a constant-speed propeller is that it converts a high percentage of brake horsepower (BHP) into thrust horsepower (THP) over a wide range of r.p.m. and airspeed combinations. A constant-speed propeller is more efficient than other propellers because it allows selection of the most efficient engine r.p.m. for the given conditions.

An airplane with a constant-speed propeller has two controls—the throttle and the propeller control. The throttle controls power output, and the propeller control regulates engine r.p.m. and, in turn, propeller r.p.m., which is registered on the tachometer.

Once a specific r.p.m. is selected, a governor automatically adjusts the propeller blade angle as necessary to maintain the selected r.p.m. For example, after setting the desired r.p.m. during cruising flight, an increase in airspeed or decrease in propeller load will cause the propeller blade angle to increase as necessary to maintain the selected r.p.m. A reduction in airspeed or increase in propeller load will cause the propeller blade angle to decrease.

The range of possible blade angles for a constant-speed propeller is the propeller′s constant-speed range and is defined by the high and low pitch stops. As long as the propeller blade angle is within the constant-speed range and not against either pitch stop, a constant engine r.p.m. will be maintained. However, once the propeller blades contact a pitch stop, the engine r.p.m. will increase or decrease as appropriate, with changes in airspeed and propeller load. For example, once a specific r.p.m. has been selected, if aircraft speed decreases enough to rotate the propeller blades until they contact the low pitch stop, any further decrease in airspeed will cause engine r.p.m. to decrease the same way as if a fixed-pitch propeller were installed. The same holds true when an airplane equipped with a constant-speed propeller accelerates to a faster airspeed. As the aircraft accelerates, the propeller blade angle increases to maintain the selected r.p.m. until the high pitch stop is reached. Once this occurs, the blade angle cannot increase any further and engine r.p.m. increases.

On airplanes that are equipped with a constant-speed propeller, power output is controlled by the throttle and indicated by a manifold pressure gauge. The gauge measures the absolute pressure of the fuel/air mixture inside the intake manifold and is more correctly a measure of manifold absolute pressure (MAP). At a constant r.p.m. and altitude, the amount of power produced is directly related to the fuel/air flow being delivered to the combustion chamber. As you increase the throttle setting, more fuel and air is flowing to the engine; therefore, MAP increases. When the engine is not running, the manifold pressure gauge indicates ambient air pressure (i.e., 29.92 in. Hg). When the engine is started, the manifold pressure indication will decrease to a value less than ambient pressure (i.e., idle at 12 in. Hg). Correspondingly, engine failure or power loss is indicated on the manifold gauge as an increase in manifold pressure to a value corresponding to the ambient air pressure at the altitude where the failure occurred.

Engine power output is indicated on the manifold pressure gauge.

The manifold pressure gauge is color-coded to indicate the engine´s operating range. The face of the manifold pressure gauge contains a green arc to show the normal operating range, and a red radial line to indicate the upper limit of manifold pressure.

For any given r.p.m., there is a manifold pressure that should not be exceeded. If manifold pressure is excessive for a given r.p.m., the pressure within the cylinders could be exceeded, thus placing undue stress on the cylinders. If repeated too frequently, this stress could weaken the cylinder components, and eventually cause engine failure.

You can avoid conditions that could overstress the cylinders by being constantly aware of the r.p.m., especially when increasing the manifold pressure.

Conform to the manufacturer´s recommendations for power settings of a particular engine so as to maintain the proper relationship between manifold pressure and r.p.m.

When both manifold pressure and r.p.m. need to be changed, avoid engine overstress by making power adjustments in the proper order:

- When power settings are being decreased, reduce manifold pressure before reducing r.p.m. If r.p.m. is reduced before manifold pressure, manifold pressure will automatically increase and possibly exceed the manufacturer´s tolerances.
- When power settings are being increased, reverse the order—increase r.p.m. first, then manifold pressure.
- To prevent damage to radial engines, operating time at maximum r.p.m. and manifold pressure must be held to a minimum, and operation at maximum r.p.m. and low manifold pressure must be avoided.

Under normal operating conditions, the most severe wear, fatigue, and damage to high performance reciprocating engines occurs at high r.p.m. and low manifold pressure.

Induction Systems

The induction system brings in air from the outside, mixes it with fuel, and delivers the fuel/air mixture to the cylinder where combustion occurs. Outside air enters the induction system through an intake port on the front of the engine cowling. This port normally contains an air filter that inhibits the entry of dust and other foreign objects. Since the filter may occasionally become clogged, an alternate source of air must be available. Usually, the alternate air comes from inside the engine cowling, where it bypasses a clogged air filter. Some alternate air sources function automatically, while others operate manually.

Two types of induction systems are commonly used in small airplane engines:

Carburetor Systems

Carburetors are classified as either float-type or pressure-type. Pressure carburetors are usually not found on small airplanes. The basic difference between a pressure carburetor and a float-type is the pressure carburetor delivers fuel under pressure by a fuel pump.

In the operation of the float-type carburetor system, the outside air first flows through an air filter, usually located at an air intake in the front part of the engine cowling. This filtered air flows into the carburetor and through a venturi, a narrow throat in the carburetor.

When the air flows through the venturi, a low-pressure area is created, which forces the fuel to flow through a main fuel jet located at the throat. The fuel then flows into the airstream, where it is mixed with the flowing air.

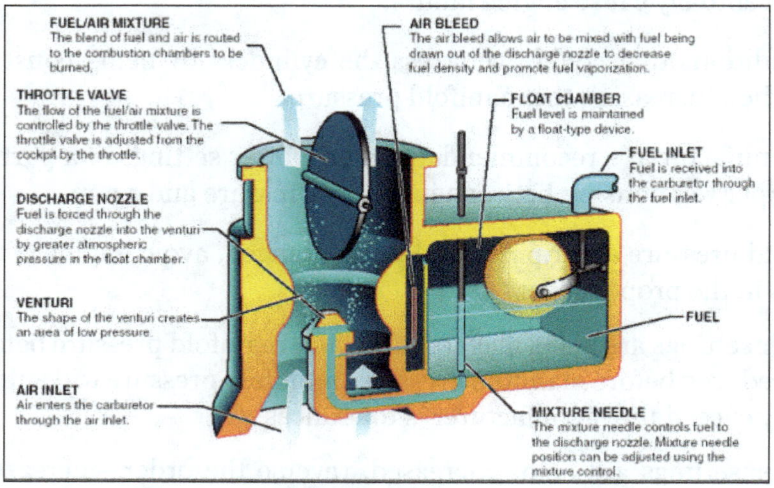

Float-type carburetor.

The fuel/air mixture is then drawn through the intake manifold and into the combustion chambers, where it is ignited. The "float-type carburetor" acquires its name from a float, which rests on fuel within the float chamber. A needle attached to the float opens and closes an opening at the bottom of the carburetor bowl.

This meters the correct amount of fuel into the carburetor, depending upon the position of the float, which is controlled by the level of fuel in the float chamber. When the level of the fuel forces the float to rise, the needle valve closes the fuel opening and shuts off the fuel flow to the carburetor. The needle valve opens again when the engine requires additional fuel.

The flow of the fuel/air mixture to the combustion chambers is regulated by the throttle valve, which is controlled by the throttle in the cockpit.

Mixture Control

Carburetors are normally calibrated at sea-level pressure, where the correct fuel-to-air mixture ratio is established with the mixture control set in the FULL RICH position. However, as altitude increases, the density of air entering the carburetor decreases, while the density of the fuel remains the same. This creates a progressively richer mixture, which can result in engine roughness and an appreciable loss of power. The roughness normally is due to spark plug fouling from excessive carbon buildup on the plugs. Carbon buildup occurs because the excessively rich mixture lowers the temperature inside the cylinder, inhibiting complete combustion of the fuel. This condition may occur during the pretakeoff runup at high-elevation airports and during climbs or cruise flight at high altitudes. To maintain the correct fuel/air mixture, you must lean the mixture using the mixture control. Leaning the mixture decreases fuel flow, which compensates for the decreased air density at high altitude.

During a descent from high altitude, the opposite is true. The mixture must be enriched, or it may become too lean. An overly lean mixture causes detonation, which may result in rough engine operation, overheating, and a loss of power. The best way to maintain the proper mixture is to monitor the engine temperature and enrichen the mixture as needed.

Proper mixture control and better fuel economy for fuel-injected engines can be achieved by use of an exhaust gas temperature gauge. Since the process of adjusting the mixture can vary from one airplane to another, it is important to refer to the Airplane Flight Manual (AFM) or the Pilot´s Operating Handbook (POH) to determine the specific procedures for a given airplane.

Carburetor Icing

One disadvantage of the float-type carburetor is its icing tendency. Carburetor ice occurs due to the effect of fuel vaporization and the decrease in air pressure in the venturi, which causes a sharp temperature drop in the carburetor. If water vapor in the air condenses when the carburetor temperature is at or below freezing, ice may form on internal surfaces of the carburetor, including the throttle valve.

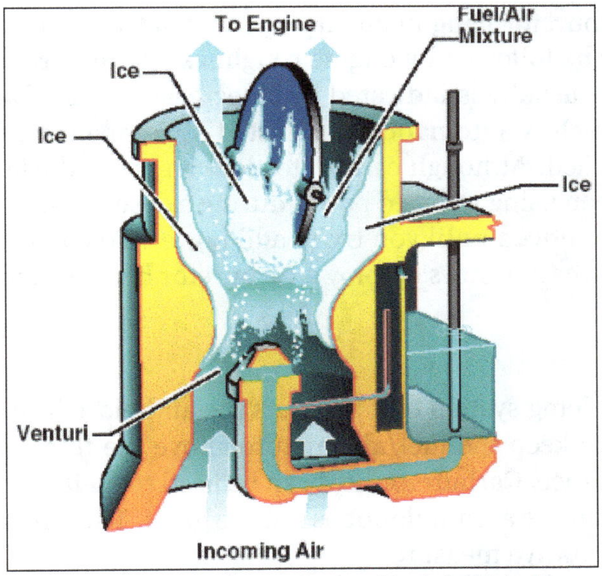

The formation of carburetor ice may reduce or block fuel/air flow to the engine.

The reduced air pressure, as well as the vaporization of fuel, contributes to the temperature decrease in the carburetor. Ice generally forms in the vicinity of the throttle valve and in the venturi throat. This restricts the flow of the fuel/air mixture and reduces power. If enough ice builds up, the engine may cease to operate.

Carburetor ice is most likely to occur when temperatures are below 70°F (21°C) and the relative humidity is above 80 percent. However, due to the sudden cooling that takes place in the carburetor, icing can occur even with temperatures as high as 100°F (38°C) and humidity as low as 50 percent. This temperature drop can be as much as 60 to 70°F.

Therefore, at an outside air temperature of 100°F, a temperature drop of 70°F results in an air temperature in the carburetor of 30°F.

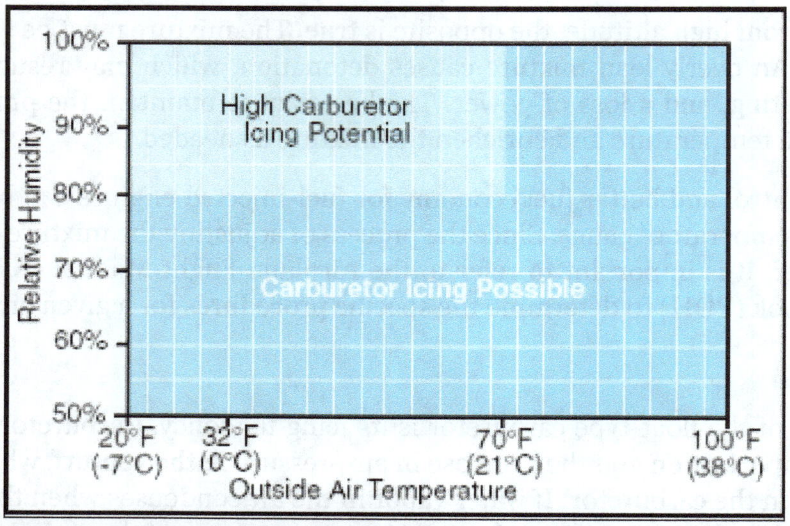

Although carburetor ice is most likely to form when the temperature and humidity are in ranges indicated by this chart, carburetor ice is possible under conditions not depicted.

The first indication of carburetor icing in an airplane with a fixed-pitch propeller is a decrease in engine r.p.m., which may be followed by engine roughness. In an airplane with a constant-speed propeller, carburetor icing usually is indicated by a decrease in manifold pressure, but no reduction in r.p.m. Propeller pitch is automatically adjusted to compensate for loss of power. Thus, a constant r.p.m. is maintained. Although carburetor ice can occur during any phase of flight, it is particularly dangerous when using reduced power during a descent. Under certain conditions, carburetor ice could build unnoticed until you try to add power. To combat the effects of carburetor ice, engines with float-type carburetors employ a carburetor heat system.

Carburetor Heat

Carburetor heat is an anti-icing system that preheats the air before it reaches the carburetor. Carburetor heat is intended to keep the fuel/air mixture above the freezing temperature to prevent the formation of carburetor ice. Carburetor heat can be used to melt ice that has already formed in the carburetor provided that the accumulation is not too great. The emphasis, however, is on using carburetor heat as a preventative measure.

The carburetor heat should be checked during the engine runup. When using carburetor heat, follow the manufacturer´s recommendations.

When conditions are conducive to carburetor icing during flight, periodic checks should be made to detect its presence. If detected, full carburetor heat should be applied immediately, and it should be left in the ON position until you are certain that all the ice has been removed. If ice is present, applying partial heat or leaving heat on for an insufficient time might aggravate the situation. In extreme cases of carburetor icing, even after the ice has been removed, full carburetor heat should be used to prevent further ice formation. A carburetor temperature gauge, if installed, is very useful in determining when to use carburetor heat.

Whenever the throttle is closed during flight, the engine cools rapidly and vaporization of the fuel is less complete than if the engine is warm. Also, in this condition, the engine is more susceptible to carburetor icing. Therefore, if you suspect carburetor icing conditions and anticipate closed-throttle operation, adjust the carburetor heat to the full ON position before closing the throttle, and leave it on during the closed-throttle operation. The heat will aid in vaporizing the fuel, and help prevent the formation of carburetor ice. Periodically, open the throttle smoothly for a few seconds to keep the engine warm, otherwise the carburetor heater may not provide enough heat to prevent icing.

The use of carburetor heat causes a decrease in engine power, sometimes up to 15 percent, because the heated air is less dense than the outside air that had been entering the engine. This enriches the mixture. When ice is present in an airplane with a fixed-pitch propeller and carburetor heat is being used, there is a decrease in r.p.m., followed by a gradual increase in r.p.m. as the ice melts. The engine also should run more smoothly after the ice has been removed. If ice is not present, the r.p.m. will decrease, then remain constant. When carburetor heat is used on an airplane with a constant-speed propeller, and ice is present, a decrease in the manifold pressure will be noticed, followed by a gradual increase. If carburetor icing is not present, the gradual increase in manifold pressure will not be apparent until the carburetor heat is turned off.

It is imperative that a pilot recognizes carburetor ice when it forms during flight. In addition, a loss of power, altitude, and/or airspeed will occur. These symptoms may sometimes be accompanied by vibration or engine roughness. Once a power loss is noticed, immediate action should be taken to eliminate ice already formed in the carburetor, and to prevent further ice formation. This is accomplished by applying full carburetor heat, which will cause a further reduction in power, and possibly engine roughness as melted ice goes through the engine. These symptoms may last from 30 seconds to several minutes, depending on the severity of the icing. During this period, the pilot must resist the temptation to decrease the carburetor heat usage. Carburetor heat must remain in the full-hot position until normal power returns.

Since the use of carburetor heat tends to reduce the output of the engine and also to increase the operating temperature, carburetor heat should not be used when full power is required (as during takeoff) or during normal engine operation, except to check for the presence or to remove carburetor ice.

Carburetor Air Temperature Gauge

Some airplanes are equipped with a carburetor air temperature gauge, which is useful in detecting potential icing conditions. Usually, the face of the gauge is calibrated in degrees Celsius (°C), with a yellow arc indicating the carburetor air temperatures where icing may occur. This yellow arc typically ranges between -15°C and +5°C (5°F and 41°F). If the air temperature and moisture content of the air are such that carburetor icing is improbable, the engine can be operated with the indicator in the yellow range with no adverse effects. However, if the atmospheric conditions are conducive to carburetor icing, the indicator must be kept outside the yellow arc by application of carburetor heat.

Certain carburetor air temperature gauges have a red radial, which indicates the maximum permissible carburetor inlet air temperature recommended by the engine manufacturer; also, a green arc may be included to indicate the normal operating range.

Outside Air Temperature Gauge

Most airplanes also are equipped with an outside air temperature (OAT) gauge calibrated in both degrees Celsius and Fahrenheit. It provides the outside or ambient air temperature for calculating true airspeed, and also is useful in detecting potential icing conditions.

Fuel Injection Systems

In a fuel injection system, the fuel is injected either directly into the cylinders, or just ahead of the intake valve. A fuel injection system is considered to be less susceptible to icing than the carburetor system. Impact icing on the air intake, however, is a possibility in either system. Impact icing occurs when ice forms on the exterior of the airplane, and blocks openings such as the air intake for the injection system.

The air intake for the fuel injection system is similar to that used in the carburetor system, with an alternate air source located within the engine cowling. This source is used if the external air source is obstructed. The alternate air source is usually operated automatically, with a backup manual system that can be used if the automatic feature malfunctions.

A fuel injection system usually incorporates these basic components—an engine-driven fuel pump, a fuel/air control unit, fuel manifold (fuel distributor), discharge nozzles, an auxiliary fuel pump, and fuel pressure/flow indicators.

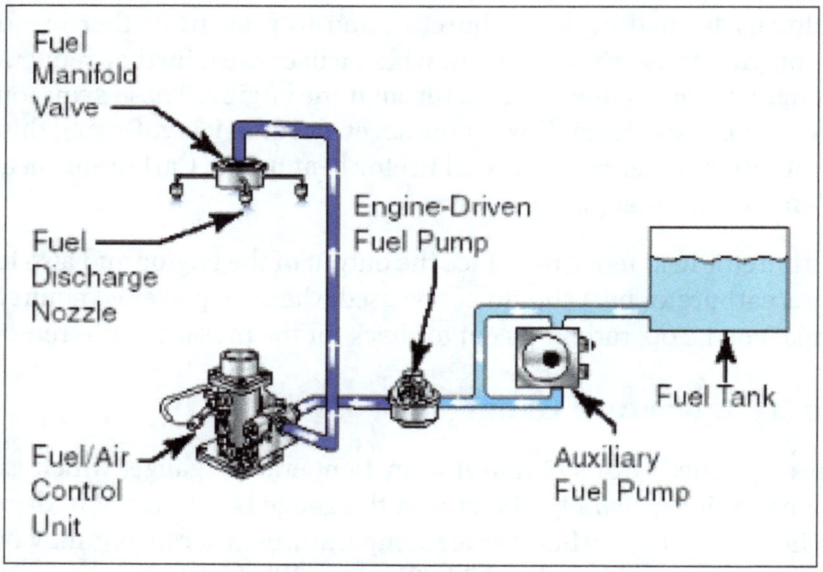

Fuel injection system.

The auxiliary fuel pump provides fuel under pressure to the fuel/air control unit for engine starting and/or emergency use. After starting, the engine-driven fuel pump provides fuel under pressure from the fuel tank to the fuel/air control unit. This control unit, which essentially replaces the carburetor, meters fuel based on the mixture control setting, and sends it to the fuel manifold valve at a rate controlled by the throttle. After reaching the fuel manifold valve, the fuel is distributed to the individual fuel discharge nozzles. The discharge nozzles, which are located in each cylinder head, inject the fuel/air mixture directly into each cylinder intake port.

Some of the advantages of fuel injection are:

- Reduction in evaporative icing.
- Better fuel flow.
- Faster throttle response.
- Precise control of mixture.
- Better fuel distribution.
- Easier cold weather starts.

Disadvantages usually include:

- Difficulty in starting a hot engine.
- Vapor locks during ground operations on hot days.
- Problems associated with restarting an engine that quits because of fuel starvation.

Superchargers and Turbosuperchargers

To increase an engine´s horsepower, manufacturers have developed supercharger and turbosupercharger systems that compress the intake air to increase its density. Airplanes with these systems have a manifold pressure gauge, which displays manifold absolute pressure (MAP) within the engine´s intake manifold.

On a standard day at sea level with the engine shut down, the manifold pressure gauge will indicate the ambient absolute air pressure of 29.92 in. Hg. Because atmospheric pressure decreases approximately 1 in. Hg per 1,000 feet of altitude increase, the manifold pressure gauge will indicate approximately 24.92 in. Hg at an airport that is 5,000 feet above sea level with standard day conditions.

As a normally aspirated aircraft climbs, it eventually reaches an altitude where the MAP is insufficient for a normal climb. That altitude limit is the aircraft´s service ceiling, and it is directly affected by the engine´s ability to produce power. If the induction air entering the engine is pressurized, or boosted, by either a supercharger or a turbosupercharger, the aircraft´s service ceiling can be increased. With these systems, you can fly at higher altitudes with the advantage of higher true airspeeds and the increased ability to circumnavigate adverse weather.

Superchargers

A supercharger is an engine-driven air pump or compressor that increases manifold pressure and forces the fuel/air mixture into the cylinders. The higher the manifold pressure, the more dense the fuel/air mixture, and the more power an engine can produce. With a normally aspirated engine, it is not possible to have manifold pressure higher than the existing atmospheric pressure. A supercharger is capable of boosting manifold pressure above 30 in. Hg.

The components in a supercharged induction system are similar to those in a normally aspirated system, with the addition of a supercharger between the fuel metering device and intake manifold.

A supercharger is driven by the engine through a gear train at one speed, two speeds, or variable speeds. In addition, superchargers can have one or more stages. Each stage provides an increase in pressure. Therefore, superchargers may be classified as single stage, two stage, or multistage, depending on the number of times compression occurs.

An early version of a single-stage, single-speed supercharger may be referred to as a sea-level supercharger. An engine equipped with this type of supercharger is called a sea-level engine. With this type of supercharger, a single gear-driven impeller is used to increase the power produced by an engine at all altitudes. The drawback, however, is that with this type of supercharger, engine power output still decreases with an increase in altitude, in the same way that it does with a normally aspirated engine.

Single-stage, single-speed superchargers are found on many high-powered radial engines, and use an air intake that faces forward so the induction system can take full advantage of the ram air. Intake air passes through ducts to a carburetor, where fuel is metered in proportion to the airflow. The fuel/air charge is then ducted to the supercharger, or blower impeller, which accelerates the fuel/air mixture outward. Once accelerated, the fuel/air mixture passes through a diffuser, where air velocity is traded for pressure energy. After compression, the resulting high pressure fuel/air mixture is directed to the cylinders.

Some of the large radial engines developed during World War II have a single-stage, two-speed supercharger. With this type of supercharger, a single impeller may be operated at two speeds. The low impeller speed is often referred to as the low blower setting, while the high impeller speed is called the high blower setting. On engines equipped with a two-speed supercharger, a lever or switch in the cockpit activates an oil-operated clutch that switches from one speed to the other.

Under normal operations, takeoff is made with the supercharger in the low blower position. In this mode, the engine performs as a ground-boosted engine, and the power output decreases as the aircraft gains altitude. However, once the aircraft reaches a specified altitude, a power reduction is made, and the supercharger control is switched to the high blower position. The throttle is then reset to the desired manifold pressure. An engine equipped with this type of supercharger is called an altitude engine.

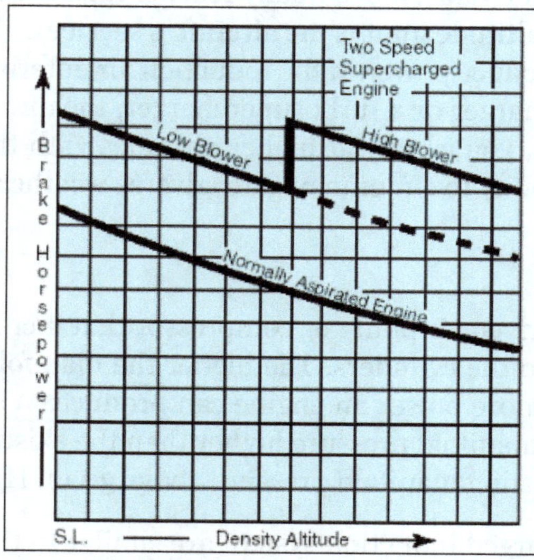

Power output of normally aspirated engine compared to a single-stage, two-speed supercharged engine.

Turbosuperchargers

The most efficient method of increasing horsepower in a reciprocating engine is by use of a turbo-supercharger, or turbocharger, as it is usually called. A drawback of gear-driven superchargers is that they use a large amount of the engine´s power output for the amount of power increase they produce. This problem is avoided with a turbocharger, because turbochargers are powered by an engine´s exhaust gases. This means a turbocharger recovers energy from hot exhaust gases that would otherwise be lost.

Another advantage of turbochargers is that they can be controlled to maintain an engine´s rated sea-level horsepower from sea level up to the engine´s critical altitude. Critical altitude is the maximum altitude at which a turbocharged engine can produce its rated horsepower. Above the critical altitude, power output begins to decrease like it does for a normally aspirated engine.

Turbochargers increase the pressure of the engine´s induction air, which allows the engine to develop sea level or greater horsepower at higher altitudes. A turbocharger is comprised of two main elements—a turbine and a compressor. The compressor section houses an impeller that turns at a high rate of speed. As induction air is drawn across the impeller blades, the impeller accelerates the air, allowing a large volume of air to be drawn into the compressor housing. The impeller´s action subsequently produces high-pressure, high-density air, which is delivered to the engine. To turn the impeller, the engine´s exhaust gases are used to drive a turbine wheel that is mounted on the opposite end of the impeller´s drive shaft. By directing different amounts of exhaust gases to flow over the turbine, more energy can be extracted, causing the impeller to deliver more compressed air to the engine. The waste gate is used to vary the mass of exhaust gas flowing into the turbine. A waste gate is essentially an adjustable butterfly valve that is installed in the exhaust system. When closed, most of the exhaust gases from the engine are forced to flow through the turbine. When open, the exhaust gases are allowed to bypass the turbine by flowing directly out through the engine´s exhaust pipe.

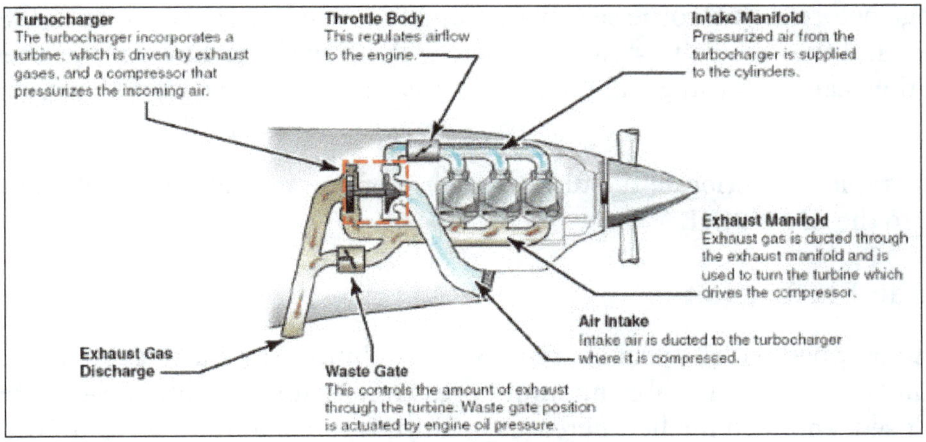

Turbocharger components

Since the temperature of a gas rises when it is compressed, turbocharging causes the temperature of the induction air to increase. To reduce this temperature and lower the risk of detonation, many turbocharged engines use an intercooler. An intercooler is a small heat exchanger that uses outside air to cool the hot compressed air before it enters the fuel metering device.

System Operation

On most modern turbocharged engines, the position of the waste gate is governed by a pressure-sensing control mechanism coupled to an actuator. Engine oil directed into or away from this actuator moves the waste gate position. On these systems, the actuator is automatically positioned to produce the desired MAP simply by changing the position of the throttle control.

Other turbocharging system designs use a separate manual control to position the waste gate. With manual control, you must closely monitor the manifold pressure gauge to determine when the desired MAP has been achieved. Manual systems are often found on aircraft that have been modified with aftermarket turbocharging systems. These systems require special operating considerations. For example, if the waste gate is left closed after descending from a high altitude, it is possible to produce a manifold pressure that exceeds the engine´s limitations. This condition is referred to as an overboost, and it may produce severe detonation because of the leaning effect resulting from increased air density during descent.

Although an automatic waste gate system is less likely to experience an overboost condition, it can still occur.

If you try to apply takeoff power while the engine oil temperature is below its normal operating range, the cold oil may not flow out of the waste gate actuator quickly enough to prevent an overboost. To help prevent overboosting, you should advance the throttle cautiously to prevent exceeding the maximum manifold pressure limits.

There are system limitations that you should be aware of when flying an aircraft with a turbocharger. For instance, a turbocharger turbine and impeller can operate at rotational speeds in excess of 80,000 r.p.m. while at extremely high temperatures. To achieve high rotational speed, the bearings within the system must be constantly supplied with engine oil to reduce the frictional forces and high temperature. To obtain adequate lubrication, the oil temperature should be in the normal operating range before high throttle settings are applied. In addition, you should allow the turbocharger to cool and the turbine to slow down before shutting the engine down. Otherwise, the oil remaining in the bearing housing will boil, causing hard carbon deposits to form on the bearings and shaft.

These deposits rapidly deteriorate the turbocharger´s efficiency and service life. For further limitations, refer to the AFM/POH.

High Altitude Performance

As an aircraft equipped with a turbocharging system climbs, the waste gate is gradually closed to maintain the maximum allowable manifold pressure. At some point, however, the waste gate will be fully closed, and with further increases in altitude, the manifold pressure will begin to decrease. This is the critical altitude, which is established by the airplane or engine manufacturer. When evaluating the performance of the turbocharging system, if the manifold pressure begins decreasing before the specified critical altitude, the engine and turbocharging system should be inspected by a qualified aviation maintenance technician to verify the system´s proper operation.

Ignition System

The ignition system provides the spark that ignites the fuel/air mixture in the cylinders and is made up of magnetos, spark plugs, high-tension leads, and the ignition switch.

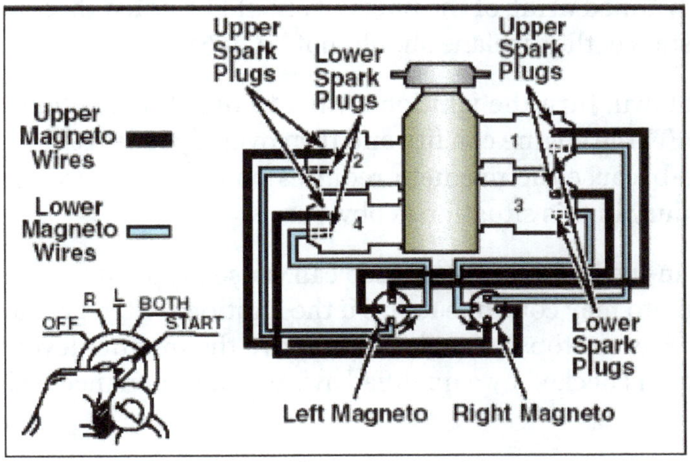

Ignition system components.

A magneto uses a permanent magnet to generate an electrical current completely independent of the aircraft´s electrical system. The magneto generates sufficiently high voltage to jump a spark across the spark plug gap in each cylinder. The system begins to fire when you engage the starter and the crankshaft begins to turn. It continues to operate whenever the crankshaft is rotating.

Most standard certificated airplanes incorporate a dual ignition system with two individual magnetos, separate sets of wires, and spark plugs to increase reliability of the ignition system. Each magneto operates independently to fire one of the two spark plugs in each cylinder. The firing of two spark plugs improves combustion of the fuel/air mixture and results in a slightly higher power output. If one of the magnetos fails, the other is unaffected. The engine will continue to operate normally, although you can expect a slight decrease in engine power. The same is true if one of the two spark plugs in a cylinder fails.

The operation of the magneto is controlled in the cockpit by the ignition switch. The switch has five positions:

1. OFF
2. R—Right
3. L—Left
4. BOTH
5. START

With RIGHT or LEFT selected, only the associated magneto is activated. The system operates on both magnetos with BOTH selected.

You can identify a malfunctioning ignition system during the pretakeoff check by observing the decrease in r.p.m. that occurs when you first move the ignition switch from BOTH to RIGHT, and then from BOTH to LEFT. A small decrease in engine r.p.m. is normal during this check. The permissible decrease is listed in the AFM or POH. If the engine stops running when you switch to one magneto or if the r.p.m. drop exceeds the allowable limit, do not fly the airplane until the problem is corrected. The cause could be fouled plugs, broken or shorted wires between the magneto and the plugs, or improperly timed firing of the plugs. It should be noted that "no drop" in r.p.m. is not normal, and in that instance, the airplane should not be flown.

Following engine shutdown, turn the ignition switch to the OFF position. Even with the battery and master switches OFF, the engine can fire and turn over if you leave the ignition switch ON and the propeller is moved because the magneto requires no outside source of electrical power. The potential for serious injury in this situation is obvious.

Loose or broken wires in the ignition system also can cause problems. For example, if the ignition switch is OFF, the magneto may continue to fire if the ignition switch ground wire is disconnected. If this occurs, the only way to stop the engine is to move the mixture lever to the idle cutoff position, then have the system checked by a qualified aviation maintenance technician.

Combustion

During normal combustion, the fuel/air mixture burns in a very controlled and predictable manner. Although the process occurs in a fraction of a second, the mixture actually begins to burn at the point where it is ignited by the spark plugs, then burns away from the plugs until it is all consumed. This type of combustion causes a smooth buildup of temperature and pressure and ensures that the expanding gases deliver the maximum force to the piston at exactly the right time in the power stroke.

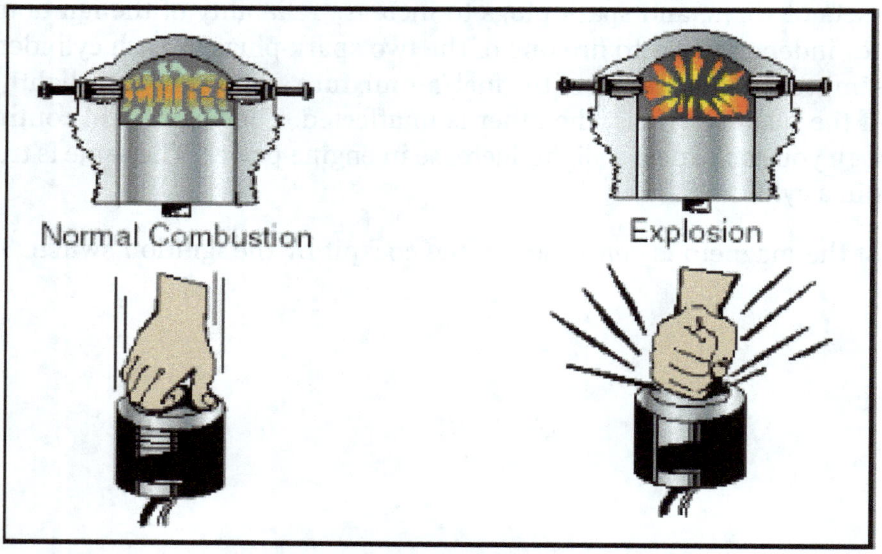

Normal combustion and explosive combustion.

Detonation is an uncontrolled, explosive ignition of the fuel/air mixture within the cylinder´s combustion chamber. It causes excessive temperatures and pressures which, if not corrected, can quickly lead to failure of the piston, cylinder, or valves. In less severe cases, detonation causes engine overheating, roughness, or loss of power.

Detonation is characterized by high cylinder head temperatures, and is most likely to occur when operating at high power settings. Some common operational causes of detonation include:

- Using a lower fuel grade than that specified by the aircraft manufacturer.
- Operating with extremely high manifold pressures in conjunction with low r.p.m.
- Operating the engine at high power settings with an excessively lean mixture.
- Detonation also can be caused by extended ground operations, or steep climbs where cylinder cooling is reduced.

Detonation may be avoided by following these basic guidelines during the various phases of ground and flight operations:

- Make sure the proper grade of fuel is being used.
- While on the ground, keep the cowl flaps (if available) in the full-open position to provide the maximum airflow through the cowling.
- During takeoff and initial climb, the onset of detonation can be reduced by using an enriched fuel mixture, as well as using a shallower climb angle to increase cylinder cooling.
- Avoid extended, high power, steep climbs.
- Develop a habit of monitoring the engine instruments to verify proper operation according to procedures established by the manufacturer.

Preignition occurs when the fuel/air mixture ignites prior to the engine´s normal ignition event. Premature burning is usually caused by a residual hot spot in the combustion chamber, often created by a small carbon deposit on a spark plug, a cracked spark plug insulator, or other damage in the cylinder that causes a part to heat sufficiently to ignite the fuel/air charge.

Preignition causes the engine to lose power, and produces high operating temperature. As with detonation, preignition may also cause severe engine damage, because the expanding gases exert excessive pressure on the piston while still on its compression stroke.

Detonation and preignition often occur simultaneously and one may cause the other. Since either condition causes high engine temperature accompanied by a decrease in engine performance, it is often difficult to distinguish between the two. Using the recommended grade of fuel and operating the engine within its proper temperature, pressure, and r.p.m. ranges reduce the chance of detonation or preignition.

Fuel Systems

The fuel system is designed to provide an uninterrupted flow of clean fuel from the fuel tanks to the engine. The fuel must be available to the engine under all conditions of engine power, altitude, attitude, and during all approved flight maneuvers. Two common classifications apply to fuel systems in small airplanes - gravity-feed and fuel-pump systems.

The gravity-feed system utilizes the force of gravity to transfer the fuel from the tanks to the engine - for example, on high-wing airplanes where the fuel tanks are installed in the wings. This places the fuel tanks above the carburetor, and the fuel is gravity fed through the system and into the carburetor. If the design of the airplane is such that gravity cannot be used to transfer fuel, fuel pumps are installed - for example, on low-wing airplanes where the fuel tanks in the wings are located below the carburetor.

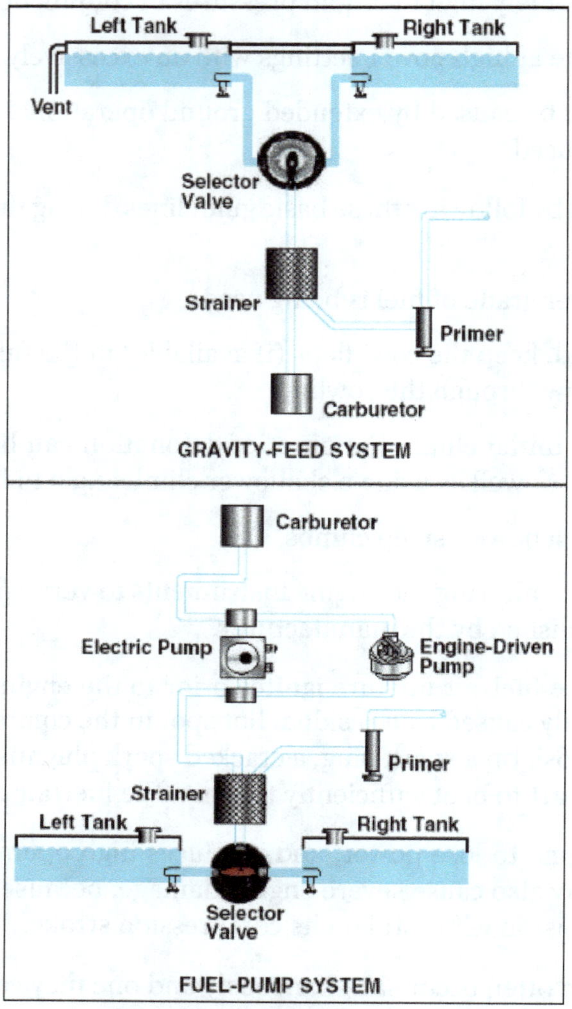

Gravity-feed and fuel-pump systems.

Fuel Pumps

Airplanes with fuel pump systems have two fuel pumps. The main pump system is engine driven, and an electrically driven auxiliary pump is provided for use in engine starting and in the event the engine pump fails. The auxiliary pump, also known as a boost pump, provides added reliability to the fuel system. The electrically driven auxiliary pump is controlled by a switch in the cockpit.

Fuel Primer

Both gravity fed and pump systems may incorporate a fuel primer into the system. The primer is used to draw fuel from the tanks to vaporize it directly into the cylinders prior to starting the engine. This is particularly helpful during cold weather, when engines are hard to start because there is not enough heat available to vaporize the fuel in the carburetor. It is important to lock the primer in place when it is not in use. If the knob is free to move, it may vibrate out during flight and can cause an excessively rich mixture. To avoid overpriming, read the priming instructions for your airplane.

Fuel Tanks

The fuel tanks, normally located inside the wings of an airplane, have a filler opening on top of the wing through which they can be filled. A filler cap covers this opening. The tanks are vented to the outside to maintain atmospheric pressure inside the tank. They may be vented through the filler cap or through a tube extending through the surface of the wing. Fuel tanks also include an overflow drain that may stand alone or be collocated with the fuel tank vent. This allows fuel to expand with increases in temperature without damage to the tank itself. If the tanks have been filled on a hot day, it is not unusual to see fuel coming from the overflow drain.

Fuel Gauges

The fuel quantity gauges indicate the amount of fuel measured by a sensing unit in each fuel tank and is displayed in gallons or pounds. Aircraft certification rules only require accuracy in fuel gauges when they read "empty." Any reading other than "empty" should be verified. Do not depend solely on the accuracy of the fuel quantity gauges. Always visually check the fuel level in each tank during the preflight inspection, and then compare it with the corresponding fuel quantity indication.

If a fuel pump is installed in the fuel system, a fuel pressure gauge is also included. This gauge indicates the pressure in the fuel lines. The normal operating pressure can be found in the AFM/POH, or on the gauge by color coding.

Fuel Selectors

The fuel selector valve allows selection of fuel from various tanks. A common type of selector valve contains four positions: LEFT, RIGHT, BOTH, and OFF. Selecting the LEFT or RIGHT position allows fuel to feed only from that tank, while selecting the BOTH position feeds fuel from both tanks. The LEFT or RIGHT position may be used to balance the amount of fuel remaining in each wing tank.

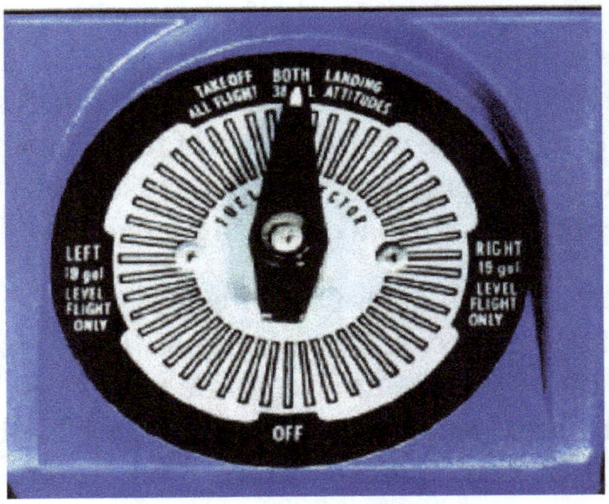

Fuel selector valve.

Fuel placards will show any limitations on fuel tank usage, such as "level flight only" and/or "both" for landings and takeoffs.

Regardless of the type of fuel selector in use, fuel consumption should be monitored closely to ensure that a tank does not run completely out of fuel. Running a fuel tank dry will not only cause the engine to stop, but running for prolonged periods on one tank causes an unbalanced fuel load between tanks. Running a tank completely dry may allow air to enter the fuel system, which may cause vapor lock. When this situation develops, it may be difficult to restart the engine. On fuel-injected engines, the fuel may become so hot it vaporizes in the fuel line, not allowing fuel to reach the cylinders.

Fuel Strainers, Sumps, and Drains

After the fuel selector valve, the fuel passes through a strainer before it enters the carburetor. This strainer removes moisture and other sediments that might be in the system. Since these contaminants are heavier than aviation fuel, they settle in a sump at the bottom of the strainer assembly. A sump is defined as a low point in a fuel system and/or fuel tank. The fuel system may contain sump, fuel strainer, and fuel tank drains, some of which may be collocated.

The fuel strainer should be drained before each flight.

Fuel samples should be drained and checked visually for water and contaminants. Water in the sump is hazardous because in cold weather the water can freeze and block fuel lines. In warm weather, it can flow into the carburetor and stop the engine. If water is present in the sump, it is likely there is more water in the fuel tanks, and you should continue to drain them until there is no evidence of water. In any event, never take off until you are certain that all water and contaminants have been removed from the engine fuel system.

Because of the variation in fuel systems, you should become thoroughly familiar with the systems that apply to your airplane. Consult the AFM or POH for specific operating procedures.

Fuel Grades

Aviation gasoline, or AVGAS, is identified by an octane or performance number (grade), which designates the antiknock value or knock resistance of the fuel mixture in the engine cylinder. The higher the grade of gasoline, the more pressure the fuel can withstand without detonating. Lower grades of fuel are used in lower-compression engines because these fuels ignite at a lower temperature. Higher grades are used in higher-compression engines, because they must ignite at higher temperatures, but not prematurely. If the proper grade of fuel is not available, use the next higher grade as a substitute. Never use a lower grade. This can cause the cylinder head temperature and engine oil temperature to exceed their normal operating range, which may result in detonation.

Several grades of aviation fuel are available. Care must be exercised to ensure that the correct aviation grade is being used for the specific type of engine. The proper fuel grade is stated in the AFM or POH, on placards in the cockpit, and next to the filler caps. Due to its lead content, auto gas should NEVER be used in aircraft engines unless the aircraft has been modified with a Supplemental Type Certificate (STC) issued by the Federal Aviation Administration.

The current method to identify aviation gasoline for aircraft with reciprocating engines is by the octane and performance number, along with the abbreviation AVGAS. These aircraft use AVGAS 80, 100, and 100LL. Although AVGAS 100LL performs the same as grade 100, the "LL" indicates it has a low lead content.

Fuel for aircraft with turbine engines is classified as JET A, JET A-1, and JET B. Jet fuel is basically kerosene and has a distinctive kerosene smell.

Since use of the correct fuel is critical, dyes are added to help identify the type and grade of fuel.

FUEL TYPE AND GRADE	COLOR OF FUEL	EQUIPMENT COLOR
AVGAS 80	RED	80 AVGAS
AVGAS 100	GREEN	100 AVGAS
AVGAS 100LL	BLUE	100LL AVGAS
JET A	COLORLESS OR STRAW	JET A

Aviation fuel color-coding system.

In addition to the color of the fuel itself, the color-coding system extends to decals and various airport fuel handling equipment. For example, all aviation gasolines are identified by name, using white letters on a red background. In contrast, turbine fuels are identified by white letters on a black background.

Fuel Contamination

Of the accidents attributed to powerplant failure from fuel contamination, most have been traced to:

- Inadequate preflight inspection by the pilot.
- Servicing aircraft with improperly filtered fuel from small tanks or drums.
- Storing aircraft with partially filled fuel tanks.
- Lack of proper maintenance.

Fuel should be drained from the fuel strainer quick drain and from each fuel tank sump into a transparent container, and then checked for dirt and water. When the fuel strainer is being drained, water in the tank may not appear until all the fuel has been drained from the lines leading to the tank. This indicates that water remains in the tank, and is not forcing the fuel out of the fuel lines leading to the fuel strainer. Therefore, drain enough fuel from the fuel strainer to be certain that fuel is being drained from the tank. The amount will depend on the length of fuel line from the tank to the drain. If water or other contaminants are found in the first sample, drain further samples until no trace appears.

Water may also remain in the fuel tanks after the drainage from the fuel strainer had ceased to show any trace of water. This residual water can be removed only by draining the fuel tank sump drains.

Water is the principal fuel contaminant. Suspended water droplets in the fuel can be identified by a cloudy appearance of the fuel or by the clear separation of water from the colored fuel, which occurs after the water has settled to the bottom of the tank. As a safety measure, the fuel sumps should be drained before every flight during the preflight inspection.

Fuel tanks should be filled after each flight, or at least after the last flight of the day to prevent moisture condensation within the tank. Another way to prevent fuel contamination is to avoid refueling from cans and drums. Refueling from cans or drums may result in fuel contamination.

The use of a funnel and chamois skin when refueling from cans or drums is hazardous under any conditions, and should be discouraged. In remote areas or in emergency situations, there may be no alternative to refueling from sources with inadequate anticontamination systems, and a chamois and funnel may be the only possible means of filtering fuel. However, the use of a chamois will not always ensure decontaminated fuel.

Worn-out chamois will not filter water; neither will a new, clean chamois that is already water-wet or damp.

Most imitation chamois skins will not filter water.

Refueling Procedures

Static electricity is formed by the friction of air passing over the surfaces of an airplane in flight and by the flow of fuel through the hose and nozzle during refueling.

Nylon, dacron, or wool clothing is especially prone to accumulate and discharge static electricity from the person to the funnel or nozzle. To guard against the possibility of static electricity igniting fuel fumes, a ground wire should be attached to the aircraft before the fuel cap is removed from the tank. The refueling nozzle then should be grounded to the aircraft before refueling is begun, and should remain grounded throughout the refueling process. When a fuel truck is used, it should be grounded prior to the fuel nozzle contacting the aircraft.

If fueling from drums or cans is necessary, proper bonding and grounding connections are important.

Drums should be placed near grounding posts, and the following sequence of connections observed:

1. Drum to ground.
1. Ground to aircraft.

2. Drum to aircraft.
3. Nozzle to aircraft before the fuel cap is removed.

When disconnecting, reverse the order.

The passage of fuel through a chamois increases the charge of static electricity and the danger of sparks. The aircraft must be properly grounded and the nozzle, chamois filter, and funnel bonded to the aircraft. If a can is used, it should be connected to either the grounding post or the funnel. Under no circumstances should a plastic bucket or similar nonconductive container be used in this operation.

Starting System

Most small aircraft use a direct-cranking electric starter system. This system consists of a source of electricity, wiring, switches, and solenoids to operate the starter and a starter motor. Most aircraft have starters that automatically engage and disengage when operated, but some older aircraft have starters that are mechanically engaged by a lever actuated by the pilot.

The starter engages the aircraft flywheel, rotating the engine at a speed that allows the engine to start and maintain operation.

Electrical power for starting is usually supplied by an on-board battery, but can also be supplied by external power through an external power receptacle. When the battery switch is turned on, electricity is supplied to the main power bus through the battery solenoid. Both the starter and the starter switch draw current from the main bus, but the starter will not operate until the starting solenoid is energized by the starter switch being turned to the "start" position. When the starter switch is released from the "start" position, the solenoid removes power from the starter motor. The starter motor is protected from being driven by the engine through a clutch in the starter drive that allows the engine to run faster than the starter motor.

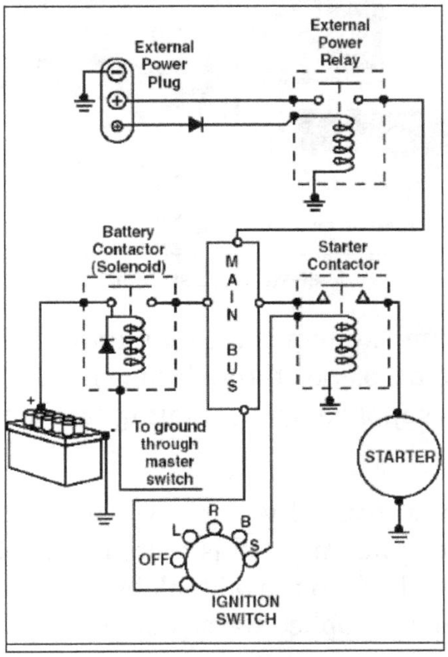

Typical starting circuit.

When starting an engine, the rules of safety and courtesy should be strictly observed. One of the most important is to make sure there is no one near the propeller. In addition, the wheels should be chocked and the brakes set, to avoid hazards caused by unintentional movement. To avoid damage to the propeller and property, the airplane should be in an area where the propeller will not stir up gravel or dust.

Oil Systems

The engine oil system performs several important functions, including:

- Lubrication of the engine´s moving parts.
- Cooling of the engine by reducing friction.
- Removing heat from the cylinders.
- Providing a seal between the cylinder walls and pistons.
- Carrying away contaminants.

Reciprocating engines use either a wet-sump or dry-sump oil system. In a dry-sump system, the oil is contained in a separate tank, and circulated through the engine by pumps. In a wet-sump system, the oil is located in a sump, which is an integral part of the engine.

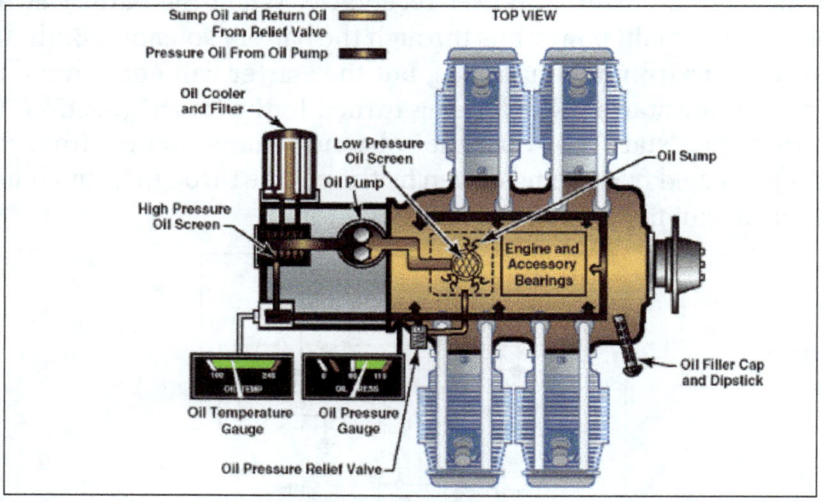

Wet-sump oil system.

The main component of a wet-sump system is the oil pump, which draws oil from the sump and routes it to the engine. After the oil passes through the engine, it returns to the sump. In some engines, additional lubrication is supplied by the rotating crankshaft, which splashes oil onto portions of the engine.

An oil pump also supplies oil pressure in a dry-sump system, but the source of the oil is a separate oil tank, located external to the engine. After oil is routed through the engine, it is pumped from the various locations in the engine back to the oil tank by scavenge pumps. Dry sump systems allow for a greater volume of oil to be supplied to the engine, which makes them more suitable for very large reciprocating engines.

The oil pressure gauge provides a direct indication of the oil system operation. It measures the pressure in pounds per square inch (p.s.i.) of the oil supplied to the engine. Green indicates the normal operating range, while red indicates the minimum and maximum pressures. There should be an indication of oil pressure during engine start. Refer to the AFM/POH for manufacturer limitations.

The oil temperature gauge measures the temperature of oil. A green area shows the normal operating range and the red line indicates the maximum allowable temperature. Unlike oil pressure, changes in oil temperature occur more slowly. This is particularly noticeable after starting a cold engine, when it may take several minutes or longer for the gauge to show any increase in oil temperature.

Check oil temperature periodically during flight especially when operating in high or low ambient air temperature. High temperature indications may indicate a plugged oil line, a low oil quantity, a blocked oil cooler, or a defective temperature gauge. Low temperature indications may indicate improper oil viscosity during cold weather operations.

The oil filler cap and dipstick (for measuring the oil quantity) are usually accessible through a panel in the engine cowling. If the quantity does not meet the manufacturer´s recommended operating levels, oil should be added. The AFM, POH, or placards near the access panel provide information about the correct oil type and weight, as well as the minimum and maximum oil quantity.

Always check the engine oil level during the preflight inspection.

Engine Cooling Systems

The burning fuel within the cylinders produces intense heat, most of which is expelled through the exhaust system. Much of the remaining heat, however, must be removed, or at least dissipated, to prevent the engine from overheating. Otherwise, the extremely high engine temperatures can lead to loss of power, excessive oil consumption, detonation, and serious engine damage.

While the oil system is vital to internal cooling of the engine, an additional method of cooling is necessary for the engine´s external surface. Most small airplanes are air cooled, although some are liquid cooled.

Air cooling is accomplished by air flowing into the engine compartment through openings in front of the engine cowling. Baffles route this air over fins attached to the engine cylinders, and other parts of the engine, where the air absorbs the engine heat. Expulsion of the hot air takes place through one or more openings in the lower, aft portion of the engine cowling.

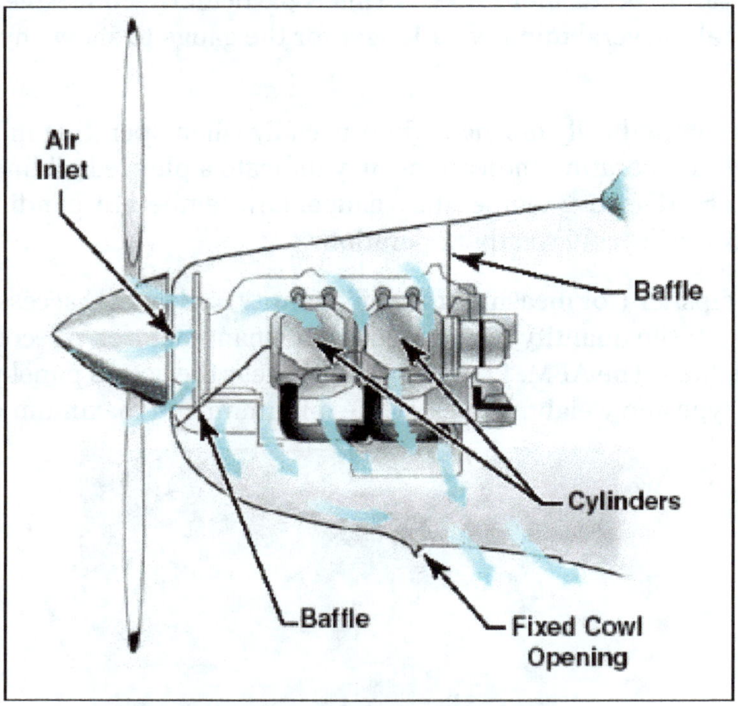

Outside air aids in cooling the engine.

The outside air enters the engine compartment through an inlet behind the propeller hub. Baffles direct it to the hottest parts of the engine, primarily the cylinders, which have fins that increase the area exposed to the airflow.

The air cooling system is less effective during ground operations, takeoffs, go-arounds, and other periods of high-power, low-airspeed operation. Conversely, high-speed descents provide excess air and can shock-cool the engine, subjecting it to abrupt temperature fluctuations.

Operating the engine at higher than its designed temperature can cause loss of power, excessive oil consumption, and detonation. It will also lead to serious permanent damage, such as scoring the cylinder walls, damaging the pistons and rings, and burning and warping the valves. Monitoring the cockpit engine temperature instruments will aid in avoiding high operating temperature.

Under normal operating conditions in airplanes not equipped with cowl flaps, the engine temperature can be controlled by changing the airspeed or the power output of the engine. High engine temperatures can be decreased by increasing the airspeed and/or reducing the power.

The oil temperature gauge gives an indirect and delayed indication of rising engine temperature, but can be used for determining engine temperature if this is the only means available.

Many airplanes are equipped with a cylinder-head temperature gauge. This instrument indicates a direct and immediate cylinder temperature change. This instrument is calibrated in degrees Celsius or Fahrenheit, and is usually color-coded with a green arc to indicate the normal operating range. A red line on the instrument indicates maximum allowable cylinder head temperature.

To avoid excessive cylinder head temperatures, increase airspeed, enrich the mixture, and/or reduce power. Any of these procedures help in reducing the engine temperature. On airplanes equipped with cowl flaps, use the cowl flap positions to control the temperature. Cowl flaps are hinged covers that fit over the opening through which the hot air is expelled. If the engine temperature is low, the cowl flaps can be closed, thereby restricting the flow of expelled hot air and increasing engine temperature. If the engine temperature is high, the cowl flaps can be opened to permit a greater flow of air through the system, thereby decreasing the engine temperature.

Exhaust Systems

Engine exhaust systems vent the burned combustion gases overboard, provide heat for the cabin, and defrost the windscreen. An exhaust system has exhaust piping attached to the cylinders, as well as a muffler and a muffler shroud. The exhaust gases are pushed out of the cylinder through the exhaust valve and then through the exhaust pipe system to the atmosphere.

For cabin heat, outside air is drawn into the air inlet and is ducted through a shroud around the muffler. The muffler is heated by the exiting exhaust gases and, in turn, heats the air around the muffler. This heated air is then ducted to the cabin for heat and defrost applications. The heat and defrost are controlled in the cockpit, and can be adjusted to the desired level.

Exhaust gases contain large amounts of carbon monoxide, which is odorless and colorless. Carbon monoxide is deadly, and its presence is virtually impossible to detect. The exhaust system must be in good condition and free of cracks.

Some exhaust systems have an exhaust gas temperature probe. This probe transmits the exhaust gas temperature (EGT) to an instrument in the cockpit.

The EGT gauge measures the temperature of the gases at the exhaust manifold. This temperature varies with the ratio of fuel to air entering the cylinders and can be used as a basis for regulating the fuel/air mixture. The EGT gauge is highly accurate in indicating the correct mixture setting. When using the EGT to aid in leaning the fuel/air mixture, fuel consumption can be reduced.

How to Become an Aircraft Powerplant Technician

Aircraft powerplant technicians are responsible for the maintenance and repair of jet and propeller engines. Individuals in this field must adhere to Federal Aviation Administration regulations, be able to work under tight deadlines and have strong mechanical skills. Read on to learn more.

Inside Aircraft Powerplant Technology

Powerplant technicians are responsible for the routine maintenance, safety inspection and repair of both jet and turboprop aircraft engines. Many specialize in a particular type of engine, aircraft inspection or a type of repair. Aircraft powerplant technicians need to complete a training program as defined and approved by the Federal Aviation Administration (FAA) before they are allowed to work on aircraft. If this occupation appeals to you, check out the following links from Study.com.

Education Information

There are about 170-approved FAA aircraft repair programs in the U.S. Powerplant technicians need to complete at least 1,900 hours of classroom instruction in a FAA-approved training program along with hands-on experience working on aircraft. Most programs may lead to an associate's or bachelor's degree, depending on training. Coursework typically includes aircraft shop practices, electrical theory, FAA regulations, fuel metering and inspection procedures. Additional information on degree options for this field are given below:

- Associate's Degree in Airframe and Powerplant Technology
- Choosing an Aircraft Powerplant School
- Choosing an Airplane Technician School
- Aircraft Powerplant Tech - Turboprop Repair Training

Career Options

There are several types of jobs to consider in aircraft repair. You could work as a repair technician, in aircraft technology or airframe repair. Below is a list of some of the occupations available.

- Aircraft Repair Technician

- Aircraft Powerplant and Airframe Repair Tech
- Aircraft Repair Tech
- Commercial Aircraft Technician

Specialization Options

You may be able to specialize in the repair of a particular part of an aircraft. Training is available in both certificate and continuing education programs. Below are a few of these options.

- Airframe and Powerplant Technology Certificate
- Certificate in Aircraft Maintenance and Training Program Overviews
- Aircraft Body and Bonded Structure Repairer

Employment Information

According to the U.S. Bureau of Labor Statistics (BLS), the number of aircraft technician jobs is expected to increase little if at all between 2012 and 2022. The latest aircraft needs less maintenance than older models and airlines may opt to outsource maintenance work as well; both of these factors contribute to the lack of significant growth for this occupation. As of May 2013, the BLS reported that the average yearly wage for aircraft mechanics and service technicians was $57,610.

Engine Repair

Engine repairs are a growing area of involvement between the FAA and applicants. Repairs are changing in their nature and scope due to many factors including part costs, extending useful engine part lives, and developing repair technologies. The FAA becomes involved in the approval of repairs to ensure that once repaired, each part meets an acceptable level of safety. Various rules, orders and guidance material focus on this activity.

What is a Repair?

A repair is the restoration of a worn or damaged engine or engine part accomplished in such a manner and using material of such quality that its restored condition will be at least equal to either its original condition, or a serviceable condition in accordance with the Instructions for Continued Airworthiness (ICAs).

How are major and minor repairs defined?

FAR 1 defines a Major repair as a repair that if improperly done, might appreciably affect weight, balance, structural strength, performance, powerplant operation, flight characteristics, or other qualities affecting airworthiness; or that is not done according to accepted practices or cannot be done by elementary operations. A minor repair is a repair other than a major repair.

Rules to Follow while Performing Engine Repair

FAR 43 Maintenance, Preventive Maintenance, Rebuilding, and Alteration, prescribes rules governing the maintenance, preventive maintenance, rebuilding, and alteration of any aircraft having a U.S. airworthiness certificate, foreign registered civil aircraft used in common carriage or carriage of mail under the provisions of Part 121, 127, or 135 and airframe, aircraft engines, propellers, appliances, and component parts of such aircraft.

Rules Governing Repair Stations

Part 145 titled, Repair Stations, prescribes the requirements for issuing repair station certificates and associated ratings to facilities for the maintenance and alteration of airframes, powerplants, propellers, or appliances, and prescribes the general operating rules for the holders of those certificates and ratings.

Must Everyone doing Repair Work have Technical Data Directly Approved by the FAA?

No, Special Federal Aviation Regulation, SFAR No. 36, applies to repair of an aircraft, airframe, aircraft engine, or propeller. The holder of an air carrier certificate or operating certificate, that operates large aircraft, and that has been issued operations specifications for operations required to be conducted in accordance with 14 CFR part 121 or part 135, may perform a major repair on a product using technical data that have not been approved by the Administrator, and approve that product for return to service, if authorized in accordance with this Special Federal Aviation Regulation.

The holder of a domestic repair station certificate, under 14 CFR part 145 may perform a major repair on an article for which it is rated, using technical data not approved by the Administrator, and approve that article for return to service, if authorized in accordance with this Special Federal Aviation Regulation (SFAR). If the certificate holder holds a rating limited to a component of a product or article, the holder may not, by virtue of this Special Federal Aviation Regulation, approve that product or article for return to service.

Repair Guidance for DER

Order 8110.37C, The DER Guidance Handbook, provides guidance, procedures, technical guidelines, and limitations of authority for Designated Engineering Representatives. It is to be used by all Aircraft Certification Directorates and Aircraft Certification Offices (ACO's), as an aid in the uniform administration of the DER program. This handbook contains repair guidance material for the DER.

Repair Related Advisory Circulars

There are several useful Repair related ACs, here are brief descriptions of three:

- AC 43-9c describes methods, procedures and practices that have been determined to be acceptable means of showing compliance with the general aviation maintenance record making and record keeping requirements of Title 14 of the Code of Federal Regulations 14 CFR part 43 and part 91.

- AC 43-12a provides information concerning preventive maintenance, who may perform it, the standards of performance applicable to it, authority for approving aircraft for return to service, and the applicable recording requirements.
- AC 43.13-1b contains methods, techniques, and practices acceptable to the Administrator for the inspection and repair of non-pressurized areas of civil aircraft, only when there are no manufacturer repair or maintenance instructions. This data generally pertains to minor repairs. The repairs identified in this AC may also be used as a basis for FAA approval for major repairs

How to Become an Aircraft Engine Specialist

Aircraft engine specialists are responsible for performing repair and maintenance work on aircraft engines that can include both airplanes as well as helicopters. Work activities for an aircraft engine specialist encompass a variety of tasks beyond the basic repair services. Duties may include testing and inspection, system adjustments, rebuilding and diagnostic work. Required knowledge and skills for a position as an aircraft engine specialist will vary depending on whether you are seeking an entry-level position and whether your goal is to work for a public airport or obtain a position at a military base.

Steps

1. Conduct some initial research on jobs as an aircraft engine specialist to learn more about the work to see if the field is a good match for you. Ask yourself if your goal is to work for a public airport or to get a position on a military base.

2. Evaluate your ability to work with complex machinery and to accurately diagnose and repair mechanical systems. Aircraft engine specialists must have a strong mechanical aptitude and knowledge of engineering and technology.

3. Browse the Internet and explore various programs that provide technical training in preparation for a position as an aircraft engine specialist. Be sure to check out the information available on government websites if you want to work on a military base as an aircraft engine specialist.

4. Determine what type of trade or vocational educational program that you want to enroll in to train in preparation for a position as an aircraft engine specialist. A special benefit of working for a branch of the military is the added plus of having them pay for part or all of your education in return for officially joining and training with them.

5. Arrange an appointment to speak with a recruiting officer if your career goal is to join a branch of the military and train for a position as an aircraft engine specialist. You will discover that most military positions offer individuals an excellent benefit program even while they are undergoing training.

6. Plan to visit several school campuses that offer vocational education programs focusing on aircraft engine specialization if you want to work for a public airport. Contact several instructors to get more information about your career options and to have any questions about the training requirements answered.

7. Consider more extensive training that includes business and administration coursework if you eventually want to obtain a management position and advance in your career as an aircraft engine specialist. An associate's degree or bachelor's degree will better prepare you for upper level management jobs.

8. Realize that some aircraft engine specialist positions may require that you become licensed or obtain certification in your field. Also be aware that working on a military base will likely require an extensive background check and a security clearance.

9. Apply for admission to the schools of your choice if you decide to work in a public airport or formally apply for acceptance into a training program affiliated with a branch of the military if that is your chosen path. Upon officially being accepted, make arrangements to start your education and training and look forward to a rewarding career as an aircraft engine specialist.

Personnel Authorized to Perform Inspection and Maintenance on Engines

Given they meet all applicable regulations, the holder of a powerplant certificate can perform maintenance and inspections on light-sport engines. The holder of at least a sport pilot certificate may approve any aircraft owned and operated by that pilot and issue a special Airworthiness Certificate in the light-sport category for return to service after performing preventive maintenance under the provisions of 14 CFR part 43, section 43.3 (g). All maintenance must be performed in accordance with 14 CFR part 65, section 65.81, which describes specific experience requirements and current instructions for performing maintenance.

The following is used to determine eligibility for a repairman certificate (light-sport aircraft) and appropriate rating. To be eligible for a repairman certificate (light-sport aircraft), you must:

Be at least 18 years old.

- Be able to read, speak, write, and understand English. If for medical reasons you cannot meet one of these requirements, the FAA may place limits on the repairman certificate necessary to safely perform the actions authorized by the certificate and rating.
- Demonstrate the requisite skill to determine whether a light-sport aircraft is in a condition for safe operation.
- Be a citizen of the United States, or a citizen of a foreign country who has been lawfully admitted for permanent residence in the United States.
- To be eligible for a repairman certificate (light-sport aircraft) with an inspection rating, the applicant must:
    1. Meet the requirements stated above for a repairman's certificate.
    2. Complete a 16-hour training course acceptable to the FAA on inspecting the particular class of experimental light-sport aircraft for which these privileges are intended to be exercised.
- To be eligible for a repairman certificate (light-sport aircraft) with a maintenance rating, the applicant must:
    1. Meet the requirements stated above for a repairman's certificate.
    2. Complete a training course acceptable to the FAA on maintaining the particular class of light-sport aircraft upon which the privileges are intended to be exercised. The training course must, at a minimum provide the following number of hours of instruction:
    3. For airplane class privileges: 120 hours.
    4. Weight-shift control aircraft class privileges: 104 hours.
    5. Powered parachute class privileges: 104 hours.
    6. Lighter-than-air class privileges: 80 hours.
    7. Glider class privileges: 80 hours.

The holder of a repairman certificate (light-sport aircraft) with an inspection rating may perform the annual condition inspection on a light-sport aircraft that is owned by the holder, has been issued an experimental certificate for operating a light-sport aircraft under 14 CFR part 21, section 21.191(i), and is in the same class of light-sport aircraft for which the holder has completed the training specified in the above paragraphs.

The holder of a repairman certificate (light-sport aircraft) with a maintenance rating may approve and return to service an aircraft that has been issued a special Airworthiness Certificate in the light-sport category under 14 CFR part 21, section 21.190, or any part thereof, after performing or inspecting maintenance (to include the annual condition inspection and the 100-hour inspection required by 14 CFR part 91, section 91.327), preventive maintenance, or an alteration (excluding a major repair or a major alteration on a product produced under an FAA approval). They may perform the annual condition inspection on a light-sport aircraft that has been issued an experimental certificate for operating a light-sport aircraft under 14 CFR part 21, section 21.191(i).

However, they may only perform maintenance, preventive maintenance, and an alteration on a light-sport aircraft for which the holder has completed the training specified in the preceding paragraphs. Before performing a major repair, the holder must complete additional training acceptable to the FAA and appropriate to the repair performed.

The holder of a repairman certificate (light-sport aircraft) with a maintenance rating may not approve for return to service any aircraft or part thereof unless that person has previously performed the work concerned satisfactorily. If that person has not previously performed that work, the person may show the ability to do the work by performing it under the direct supervision of a certificated and appropriately rated mechanic, or a certificated repairman who has had previous experience in the specific operation concerned. The repairman may not exercise the privileges of the certificate unless the repairman understands the current instructions of the manufacturer and the maintenance manuals for the specific operation concerned.

Authorized Personnel That Meet FAA Regulations

All applicable aviation regulatory authority regarding maintenance procedures must be met. Maintenance organizations and personnel are encouraged to contact the manufacturer for more information and guidance on any of the maintenance procedures.

It is a requirement that every individual or maintenance provider possess the required special tooling, training, or experience to perform all tasks outlined. Maintenance providers that meet the following conditions outlined below may perform engine maintenance providing they meet all of the following FAA requirements:

- Knowledge of the specific task as a result of receiving authorized training from a training provider.
- Previous experience in performing the task and formal instruction from a manufacturer's authorized training facility or "on-the-job" instruction by a manufacturer's representative.
- A suitable work environment to prevent contamination or damage to engine parts or modules is needed.
- Suitable tools and fixtures as outlined in the manufacturers' Maintenance Manual should be used while performing maintenance requiring such tooling.
- Reasonable and prudent maintenance practices should be utilized.

How to Maintain Light- Sport Rotax Engines

Some specific maintenance practices that differ from conventional certified engines is covered for background and educational acquaintance purposes only. Always refer to the current manufacturer's information when performing maintenance on any engine.

Safety regulations must be adhered to ensure maintenance personnel safety when performing maintenance and service work on any engine installation. The following information should be followed while performing maintenance.

The ignition should be off and the ignition system grounded with the battery disconnected. Secure the engine against unintentional operation. During maintenance work that requires ignition on and battery connected, secure the propeller against unintentional turning by hand, and secure and observe a propeller safety zone. This precautionary measure serves to avoid any injuries in case of an unintentional start of the engine, which can result in injuries or death. Remember, as long as the ground-cable (plead) is not properly connected to ground, the ignition is switched ON (hot).

Prevent contamination, such as metal chips, foreign material, and/or dirt, from entering the cooling, lubricating, and fuel system during maintenance. Severe burns and scalds may result if the engine is not allowed to cool down to outside air temperature before starting any work. Before reusing disassembled parts, clean with a suitable cleaning agent, check, and refit per instructions. Before every re-assembly, check for missing components. Only use adhesives, lubricants, cleaning agents, and solvents listed for use in the maintenance instructions. Observe the tightening torques for screws and nuts; overtorque or too loose connection could cause serious engine damage or failure.

The following are some general maintenance practices that provide for safety and good technique:

- Work only in a non-smoking area and not close to sparks or open flames.
- Always use the specified tools.
- During disassembling/reassembling the removal of any safety items (e.g., safety wiring, self-locking fastener) must always be followed with replacement of a new one.
- Once loosened, always replace self-securing (locking) nuts.
- Use clean screws and nuts only and inspect face of nuts and thread for damage.
- Check the contact faces and threads for damage and replace if any damaged is detected.
- At reassembly of the engine, replace all sealing rings, gaskets, securing elements, O-rings, and oil seals.
- At disassembly of the engine, mark the engine's components as necessary to provide for locating the original position of the part.
- Parts should be replaced in the same position upon reassembly.
- Any used components have wear patterns that should be replaced or matched if reused. Ensure that these marks are not erased or washed off.

To perform maintenance, the technician must follow the manufacturer's instructions. Obtain, read, and understand the information pertaining to servicing of the light-sport or experimental engine.

How to Maintain Light-Sport Jabiru Engines

Engine and Engine Compartment Inspection

Check for oil, fuel exhaust, and induction leaks and clean the entire engine and compartment before inspection. Check flywheel screw tensions to 24 foot pounds. Check the carburetor air filter and clean it by removing it from the intake housing and blowing compressed air against the direction of the intake flow. For operation in heavy dust conditions, clean air filter at shorter intervals than recommended for normal conditions. A clogged filter reduces engine performance, as well as promotes premature engine wear. The engine baffles and air ducts should be checked for condition and functionality.

Two methods can be used to check the cylinders compression. The compression gauge method is used to measure compression using a compression tracer. Readings are taken with a fully open throttle valve at engine oil temperature between 30 °C and 70 °C (90 °F to 160 °F). If readings are below 6 bar (90 psi) a check of the pistons, cylinders, valves, and cylinder heads must be undertaken.

The second method uses the pressure differential test. Check using a maximum allowable pressure loss is 25 percent. As an alternative to a compression test, a pressure differential test (leak down) can be accomplished. This is a much better test of the condition of rings, bore, head sealing, and valves. This is the normal test used in aviation and requires specific equipment. The test is carried out with the engine in warm to hot condition. Input pressure is best set at 80 psi; a second gauge reads the differential. This is done with piston on TDC on the firing stroke.

The propeller needs to be restrained. A differential of lower than 80/60 (generally a 25 percent loss) indicates a problem.

Problems can be better identified by observing where air is escaping from the cylinder, blow-by. Some examples are as follows:

1. Blow-by through the crankcase vent indicates worn rings or bore.
2. Leaking from carby indicates a poor intake valve seal.
3. Leaking from exhaust indicates a poor exhaust valve seal.
4. Head leak indicates poor head to cylinder seal.

With the problem identified, the malfunction can then be corrected. Poor compression can be an indication of a serious problem. For example, continued operation with poor compression due to a poorly sealing valve can lead to eventual valve failure and heavy damage to the piston, connecting-rod, barrel, and head.

Lubrication System

The oil should be changed as required by the manufacturer. When changed, the oil filter should also be changed. Change the oil filter at every 50 hourly inspection. Drain the oil while engine is still warm and visually check for leaks. Fill the engine with oil (approximately 2.3 liters) and check oil level. Never exceed the maximum mark. Use only registered brand oils meeting the correct specifications. Do not drain the oil cooler during a normal oil change. The cooler holds only a small amount of old oil that has negligible effect on the new oil. Taking the hoses on and off the cooler can prematurely age the oil lines and lead to hoses slipping off the cooler.

Carburetor Adjustment and Checks

To adjust the engine's idle speed, adjust the idle stop screw (7 mm screw) against throttle lever. Standard idle mixture screw position is 1¼ turns out from the seated position. Fine adjustment may be necessary to give a smooth idle. The mixture is set by selecting jet sizes. As supplied, the engine has jets to suit a majority of installations; however, the mixture may be affected by operation with a propeller that does not meet the requirements listed in the installation manual or by ambient temperature extremes. If an engine is to be used in these situations, an exhaust gas temperature (EGT) gauge should be fitted and monitored against the limits specified above. Do not change carburetor settings if EGT readings fall outside the range given without consulting with Jabiru Aircraft or our local authorized representative. The carburetor automatically adjusts the mixture to account for altitude. Visual inspection should include checks for carburetor joint degradation and carburetor linkage for full and free movement, correct positioning of stops and security.

Spark Plugs

When plugs are removed from a warm engine, the inspection of the tip of the spark plug can be used to indicate the health of the engine. If the tip of the plug is a light brown color, the plug is operating correctly. A black velvet, sooty looking plug tip generally is an indication of an overly rich mixture (check the choke and the air filter and intake). If the firing end tip is covered with oil, it is an indication of too much oil in the combustion chamber (check for worn piston rings and cylinder walls). When servicing the spark plugs, do not use steel or brass brushes for cleaning and never sandblast plugs. Clean the spark plugs with plastic brush in a solvent. Check electrode gap and, if necessary, adjust to 0.55–0.6mm (0.022 in–0.024 in) by carefully bending the electrode. Use the recommended Plugs (NGK D9EA) and place a suitable anti-seize compound on threads of the plug before installing them in the engine. Tighten spark plugs when the engine is cold and adjust engine to the correct torque value. Reconnect the ignition lead.

Exhaust System

Visually check the exhaust system for security of mounting, damage, rubbing, leaks, and general condition. Check nuts and bolts for tightness and condition; re-torque and replace if necessary.

Head Bolts

Check the head bolt torque after five hours of operation, and again after ten hours of operation. The bolts should, thereafter, be checked annually. Head bolts torque when cold to 20 ft/lb.

Tachometer and Sender

Many apparent engine problems can be caused through inaccurate tachometers. Where engine performance is observed to be outside limits, the tachometer should be checked against a calibrated instrument. Tachometer sender gap is 0.4mm (0.016 inches). The sender must have at least 60 percent covered by the tags fitted to the gearbox side of the flywheel. Ensure both tags are equal distance from sender.

Engine Inspection Charts

Propeller	Engine and Engine Compartment
Spinner * *	Check flywheel screw tensions to 24 foot pounds*
Spinner flange * *	Carburetor air filter * *
Spinner screws * *	Engine baffles and air ducts *
Propeller * *	Cylinders *
Propeller bolts/nuts - Tension *	Crankcase & front crankcase seal *
Spinner/prop tracking * *	Hoses, lines and fittings * *
	Intake and exhaust systems *
	Ignition harness, distributor caps & rotors *
	NOTE: Check for oil, fuel exhaust and induction leaks, then clean entire engine and compartment before inspection.
Annual Inspection** Each 100 Hours*	

Figure: Engine inspection charts

How to Perform Maintenance of a Jet Engine

Step 1: Defining the work scope

The first step is the arrival of a "dirty engine" at the Schiphol Airport facility. This eco-friendly building is named after Sir Frank Whittle, the British pilot who designed and patented the first turbo jet engine in 1930. Then it's time to roll up their sleeves.

Each engine is identified by a six-digit serial number and has its own maintenance plan and complete history data. There are all kinds of reasons why an engine might be replaced: deteriorating performance, time on-wing or, most commonly, because it's reached its cycle limit.

When this cycle limit is reached (e.g. 3,000 cycles), the engine will be removed and serviced. The work package level is then discussed in detail with engineering specialists, the production team and, of course, thier customer. Other airlines, besides KLM, make use of thier engine maintenance expertise.

The result is an overview of all the work that needs to be done. This is known as the "workscope".

Step 2: Cleaning & inspection

A mega scrub & go, just like in a wellness centre, involves several baths with different liquids to ensure all the dirt and oil is removed before the visual inspections begin. One of these is a borescope inspection in which a camera is used to carry out an internal examination of the engine. This is the same technique used in hospitals to examine the inside of the human body.

A very small camera attached to a flexible pipe is inserted into the engine through small holes distributed all around the engine. This enables the borescope operator to see right into the heart of the engine and view any irregularities, such as dents or cracks, on a large display. Nothing escapes attention using this technology.

Step 3: "Disassembly"

Components fall into two distinct categories: static parts and rotating parts. Rotating parts are located in the sections where large forces are generated, for example, in the compressor and turbine section. They are, obviously, subjected to more wear and tear than static parts. Just think of your car's gearbox: constant rotational movement means it wears out faster than other parts.

At a major overhaul, the engine is almost completely dismantled (about 40,000 parts). These parts are inspected and some of them are sent to the engine manufacturer or to other KLM and Air France departments, who are specialised in different parts. There, the mechanics examine and repair the parts, according to the repair manuals, to make them fully functional again.

And who says engineering is a man's world? Not in thier shop. They have female mechanics on their team.

Step 4: Repair & replace

Engine parts are very costly. Life-limited parts at the end of their life cycle have to be replaced with new ones. Some parts can be repaired at their top-notch repair shop, using robots and innovative techniques such as laser cladding or e-beam welding.

Are you wondering what the most expensive part of an engine is? Well, in an engine type CF6-80E that would be the Forward Outer Seal. This component costs USD 1.4 million, and no, this is not a typo. This is equivalent to about five Ferraris (type 458 for the experts) or an ultra-luxury villa with staff and pool. It really is an amazing amount of money for a single engine part!

Step 5: Assembling & testing

This is the phase in which they rebuild the engine. All the parts are collected and carefully assembled. Then the engine goes into the test cell. This is an insulated bunker where the engine is subjected to thorough testing under the toughest of conditions and all kinds of circumstances.

If the engine passes the tests, a certified inspector carries out a final inspection and issues a Certificate of Airworthiness.

Now the jet engine has been completely overhauled, the counter is back to zero and it's ready to fly again for years.

How to Maintain Propeller

Thinking about the forces at work on an aircraft while it's flying, we naturally think about lift, thrust, drag, etc. After all, the foundation of flight training begins with a focus on how these forces keep the aircraft in the air.

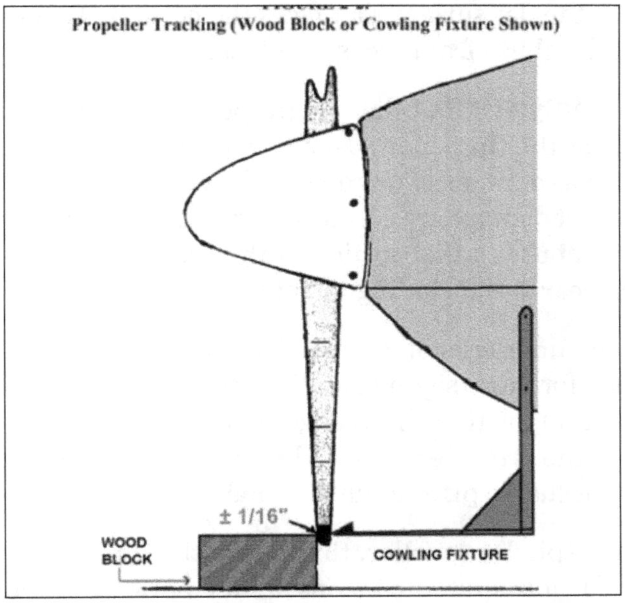

Figure: of Advisory Circular 20-37F illustrating a test of propeller tracking.

That's great from the left seat looking forward, but from a "mechanic's perspective," we tend to focus on the mechanical forces at work on the aircraft. Nothing rivals the forces placed on your propeller.

Think of it this way: Your average general aviation aircraft produces about 200 pounds to 300 pounds of thrust to fly about 130 knots, and the structure supports from -1.52 to +3.8 times the gross weight of the aircraft when operating within the limits of the normal category. Those numbers are downright quaint when compared with the forces within which the propeller operates.

The typical GA propeller experiences between 10 tons and 25 tons of centrifugal force pulling the blades from the hub of the propeller during normal operation. The blades bend and flex as they claw the air from both the thrust and torque loads, and they also have to contend with vibration from the cyclic engine power pulses and other vibrations. And most of the sound that the airport neighbors complain about comes from the propeller tips as they travel at supersonic speeds during high-rpm takeoff operations.

Given the extreme forces at work, it should come as no surprise that it is imperative that propellers be in top condition in order to operate properly. Corrosion, cracks, wear, and other issues can compromise the ability of the propeller to meet its design specifications and, ultimately, cause catastrophic failure during flight. This is serious business.

Your best defense is a good offense in the form of routine preflight, postflight, and maintenance inspections of your prop. I'm particularly fond of postflight inspections because, by definition, they give you time to plan and execute any required maintenance without get-there-itis influencing your decision-making process.

Begin your inspection by running your bare hands over the leading edges of the propeller blades, feeling for rough spots and sharp edges. Look closely for damage on all sides of the propeller, but pay particular attention to the leading edge and face of the propeller as you get closer to the tips. The face of the propeller is the flat surface that faces the cockpit. This is the side of the propeller that tends to take most of the abuse from rocks, sand, etc.

According to the FAA's AC 20-37E, "FAA data on propeller failures indicates that the majority of failures occur in the blade at the tip region, usually within several inches from the tip and often due to a crack initiator such as a pit, nick, or gouge." The combination of thickness, speed, flexing, and exposure to foreign object damage are greatest toward the tips of the propeller. There just isn't much margin for error and abuse in that section of the prop when compared to the more robust and largely protected area near and including the hub.

Most propellers experience minor erosion on the blade face and leading edge during normal use. Our main concern is to look for more significant dings that show some form of deformation of the metal in the blade and anything with a sharp edge that can serve as a "stress riser." This type of damage can cause cracks in the propeller, and it's best to find them post-flight so they can be evaluated and addressed by a mechanic prior to further flight.

If the aircraft has a variable-pitch propeller, this is also the time to look for leaks and to gently flex each blade to check for looseness or inconsistency between the different blades. A small amount of movement is normal, but defer to your mechanic if you find inconsistency or have questions.

There's another very important check that anyone can do to see if the propeller is bent or asymmetrically deformed in some way. It's called a blade tracking check. Begin by rotating the propeller until one of the blades is pointing straight down. Now place a block of wood directly under the blade and mark the exact path of the blade on the wood. Leave the wood block in place and rotate the propeller to the next blade. Each blade should align exactly with the mark on the block. If the blade passes at a different location, it indicates that the blade has been bent in some way and should be attended to immediately by a propeller shop. This type of damage can come from moving the aircraft using the propeller as a hand-hold, rather than a tow bar. It's a high price to pay for taking unnecessary shortcuts on the tarmac.

Most propeller issues fall into two categories: corrosion and physical damage.

Corrosion

One of the most insidious causes of propeller damage is corrosion. This includes external corrosion that is visible on the blades, as well as internal corrosion that literally eats away at the components within the hub of a variable-pitch propeller. Regardless of its location, corrosion reduces the structural integrity of the propeller as well as its performance.

There are three distinct types of corrosion relative to propellers: surface corrosion, pitting, and intergranular corrosion. Surface corrosion occurs when the protective coating on the propeller has been removed from the face and leading edge of the propeller, most often by the effects of rain, sand, etc. This is fairly common and can be removed by a mechanic with emery or crocus cloth, followed by re-painting of the propeller as required.

Pitting is a very specific and serious form of corrosion. Pitting corrosion consists of small, visible corrosion cavities extending inward from the metal surface of the propeller. Pits can grow on the surface, but are more commonly found where moisture is trapped, such as under decals left in place after propeller balancing or improperly installed de-icing boots. Any sign of pitting is serious and should be evaluated quickly in an effort to save the propeller from the junk heap.

Intergranular corrosion is a form of corrosion internal to the metal structure itself. It is less common and more likely to be caused by a problem in the metal casting, but it too can be caused by trapped moisture under decals or around bolt holes.

Physical Damage

The most common forms of physical propeller damage are nicks, dings, and cracks on both the blades and the hub of the propeller. A propeller shop has the tools to do much more detailed inspections of propellers for cracks, including ultrasonic, eddy current, dye penetrant, and magnetic particle inspections. However, routine inspections for damage visible to the naked eye are a crucial starting point.

According to the FAA's Advisory Circular AC 20-37E, "Limited minor repairs may be made on propellers by appropriately rated maintenance technicians either on the aircraft or when the propeller is removed. Minor dents, cuts, scars, scratches, and nicks may be removed providing their removal does not weaken the blade, substantially change weight or balance, or otherwise impair its performance."

It's important to always use the propeller manufacturer's maintenance manuals for proper maintenance procedures and limits when working on a propeller. That said, the FAA provides some guidelines for minor repairs in AC 20-37E:

For nicks, dents, pits, and cuts in the leading or trailing edges of blades, ensure that the bottom of the damage is removed first by rounding out and fairing in the repair only slightly deeper than the damage. Initial removal of material should be done using a fine cut file. All traces of file marks in the repaired area should be removed with number 240 emery cloth followed by polishing with number 320 emery cloth, then finished with crocus cloth or 600 grit emery cloth, and then visually inspected. An individual edge repair should not exceed a depth of 3/16-inch. The repair length should be 10 times longer than the depth of the repair.

For gouges, cuts, and small dents on blade faces, ensure that the bottom of the damage is removed first by rounding out and fairing in the repair to form a saucer-shaped depression only slightly deeper than the damage. The initial repair should be accomplished by filing with a fine cut file parallel to the damage and finishing with 240 and 320 emery abrasive cloth, as in the manner of damage removal from blade-leading edges. Final polishing of the repair should be done with crocus cloth or 600 grit emery cloth. An individual repair should not exceed 1/16-inch in depth and the surface radius of curvature of the repair must not be less than 3/8-inch. Repair width should be 30 times the repair depth.

Of particular note in these repair guidelines are the rules regarding the transitions and fairing out of the repairs. For leading edge repairs, the total repair length is 10 times longer than the depth of the repair. This is very important. It means that if you need to down 1/8-inch (0.125 inch) to get to the base of a nick, you will need to smooth out the repair over the distance of 1.25 inches. For blade face repairs, it's 30 times. For a 1/16-inch deep repair (0.0625 of an inch), that's a 1.875-inch disk of repair area. Those are pretty big dimensions; more than one might assume without proper training. So, it pays to know proper maintenance practices such as this to ensure that repairs are made properly.

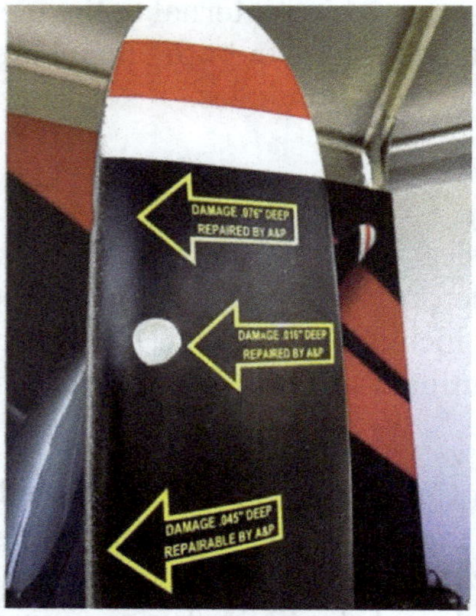

Permissions

All chapters in this book are published with permission under the Creative Commons Attribution Share Alike License or equivalent. Every chapter published in this book has been scrutinized by our experts. Their significance has been extensively debated. The topics covered herein carry significant information for a comprehensive understanding. They may even be implemented as practical applications or may be referred to as a beginning point for further studies.

We would like to thank the editorial team for lending their expertise to make the book truly unique. They have played a crucial role in the development of this book. Without their invaluable contributions this book wouldn't have been possible. They have made vital efforts to compile up to date information on the varied aspects of this subject to make this book a valuable addition to the collection of many professionals and students.

This book was conceptualized with the vision of imparting up-to-date and integrated information in this field. To ensure the same, a matchless editorial board was set up. Every individual on the board went through rigorous rounds of assessment to prove their worth. After which they invested a large part of their time researching and compiling the most relevant data for our readers.

The editorial board has been involved in producing this book since its inception. They have spent rigorous hours researching and exploring the diverse topics which have resulted in the successful publishing of this book. They have passed on their knowledge of decades through this book. To expedite this challenging task, the publisher supported the team at every step. A small team of assistant editors was also appointed to further simplify the editing procedure and attain best results for the readers.

Apart from the editorial board, the designing team has also invested a significant amount of their time in understanding the subject and creating the most relevant covers. They scrutinized every image to scout for the most suitable representation of the subject and create an appropriate cover for the book.

The publishing team has been an ardent support to the editorial, designing and production team. Their endless efforts to recruit the best for this project, has resulted in the accomplishment of this book. They are a veteran in the field of academics and their pool of knowledge is as vast as their experience in printing. Their expertise and guidance has proved useful at every step. Their uncompromising quality standards have made this book an exceptional effort. Their encouragement from time to time has been an inspiration for everyone.

The publisher and the editorial board hope that this book will prove to be a valuable piece of knowledge for students, practitioners and scholars across the globe.

Index

www.ingramcontent.com/pod-product-compliance
Lightning Source LLC
LaVergne TN
LVHW080646210926
848485LV00003B/337